T0183936

Lecture Notes in Computer Science 9664

Commenced Publication in 1973
Founding and Former Series Editors:
Gerhard Goos, Juris Hartmanis, and Jan van Leeuwen

Matthew Cook · Turlough Neary (Eds.)

Cellular Automata and Discrete Complex Systems

22nd IFIP WG 1.5 International Workshop, AUTOMATA 2016
Zurich, Switzerland, June 15–17, 2016
Proceedings

 Springer

Editors
Matthew Cook
University of Zurich
Zurich
Switzerland

Turlough Neary
University of Zurich
Zurich
Switzerland

ISSN 0302-9743   ISSN 1611-3349   (electronic)
Lecture Notes in Computer Science
ISBN 978-3-319-39299-8   ISBN 978-3-319-39300-1   (eBook)
DOI 10.1007/978-3-319-39300-1

Library of Congress Control Number: 2016939364

LNCS Sublibrary: SL1 – Theoretical Computer Science and General Issues

This Springer imprint is published by Springer Nature
The registered company is Springer International Publishing AG Switzerland

Preface

This volume contains the papers presented at the 22nd International Workshop on Cellular Automata and Discrete Complex Systems, AUTOMATA 2016. The conference was held at the University of Zurich, Switzerland, June 15–17, 2015. AUTOMATA 2016 was an IFIP event of IFIP Working Group 1.5.

The AUTOMATA series began in Dagstuhl, Germany, in 1995 and has since been run yearly, appearing in Europe, North and South America, and Asia. As with previous editions of the series, AUTOMATA 2016 provided a focal point for researchers interested in cellular automata and related discrete complex systems where they could exchange ideas, present their latest research, and forge new collaborative partnerships. The AUTOMATA series is concerned with research on all fundamental aspects of cellular automata and related discrete complex systems. Topics of interest include: dynamics, topological, ergodic and algebraic aspects, algorithmic and complexity issues, emergent properties, formal language processing, symbolic dynamics, models of parallelism and distributed systems, timing schemes, phenomenological descriptions, scientific modeling, and practical applications.

We would like to express our appreciation to the invited speakers, Klaus Sutner, Guillaume Theyssier, Tommaso Toffoli, and Andrew Winslow. We also thank them for their contributions to this volume.

We thank all the authors who submitted to AUTOMATA 2016. There were 23 submissions in total. Each paper was assigned to three members of the Program Committee and following a complete reviewing process and subsequent discussion by the committee 12 papers were accepted for publication in these proceedings. In addition, there were a number of exploratory papers presented at the conference and we would like to take this opportunity to thank those who submitted to the exploratory track.

We thank the Program Committee for their diligent work, and the reviewers who assisted in the evaluation of papers. We would like to thank the AUTOMATA Steering Committee for giving us the opportunity to host AUTOMATA 2016 at the University of Zurich and for their invaluable advice and support with a special thank you to Jarkko Kari and Pedro de Oliveira. We are grateful to the Lecture Notes in Computer Science team of Springer for all their help in preparing this volume. Finally, we would like to thank iniForum and the administrative staff at the Institute of Neuroinformatics for their help in organizing this event, in particular Kathrin Aguilar Ruiz-Hofacker, Maik Berchten, David Lawrence, and Simone Schumacher, who were always ready to provide advice and assistance.

April 2016

Turlough Neary
Matthew Cook

Organization

Program Committee

Matthew Cook	University of Zurich and ETH Zurich, Switzerland
Pedro de Oliveira	Universidade Presbiteriana Mackenzie, Brazil
Nazim Fatès	Inria Nancy Grand Est, France
Paola Flocchini	University of Ottawa, Canada
Enrico Formenti	Université Nice Sophia Antipolis, France
Anahí Gajardo	University of Concepción, Chile
Eric Goles	Adolfo Ibáñez University, Chile
Jarkko Kari	University of Turku, Finland
Martin Kutrib	Universität Giessen, Germany
Andreas Malcher	Universität Giessen, Germany
Genaro Martínez	University of the West of England, UK
Kenichi Morita	Hiroshima University, Japan
Turlough Neary	University of Zurich and ETH Zurich, Switzerland
Nicolas Ollinger	Université d'Orléans, France
Matthew Patitz	University of Arkansas, USA
Ivan Rapaport	Universidad de Chile, Chile
Hiroshi Umeo	Osaka Electro-Communication University, Japan
Damien Woods	California Institute of Technology, USA
Thomas Worsch	Karlsruhe Institute of Technology, Germany

Additional Reviewers

Silvio Capobianco	Pablo Saez
Jacob Hendricks	Georgios Ch. Sirakoulis
Rolf Hoffmann	Michal Szabados
Markus Holzer	Rodrigo Torres
Gaétan Richard	Matthias Wendlandt
Trent Rogers	Andrew Winslow

Organizing Committee

Matthew Cook	University of Zurich and ETH Zurich, Switzerland
Turlough Neary	University of Zurich and ETH Zurich, Switzerland

Steering Committee

Pedro de Oliveira	Universidade Presbiteriana Mackenzie, Brazil
Teijiro Isokawa	University of Hyogo, Japan
Jarkko Kari	University of Turku, Finland
Andreas Malcher	Universität Giessen, Germany
Thomas Worsch	Karlsruhe Institute of Technology, Germany

Sponsoring Institutions

Institute of Neuroinformatics, University of Zurich
University of Zurich

Invited Talk Abstracts

Automata, Semigroups and Dynamical Systems

Klaus Sutner

Carnegie Mellon University
Pittsburgh, PA 15213, USA

Abstract. Invertible transducers are Mealy automata that determine injective maps on Cantor space and afford compact descriptions of their associated transduction semigroups and groups. A lot of information has been unearthed in the last two decades about algebraic aspects of these machines, but relatively little is known about their automata-theoretic properties and the numerous computational problems associated with them. For example, it is quite difficult to pin down the computational complexity of the orbits of maps defined by invertible transducers. We study some of these properties in the context of so-called m-lattices, where the corresponding transduction semigroup is a free Abelian group of finite rank. In particular we show that it is decidable whether an invertible transducer belongs to this class.

Propagation, Diffusion and Randomization in Cellular Automata

Guillaume Theyssier[(✉)]

Insitut de Mathématiques de Marseille (CNRS, AMU, Centrale Marseille),
Marseille, France
guillaume.theyssier@cnrs.fr

Abstract. A large part of the study of of cellular automata dynamics consists in comparing pairs of configurations and their orbits. Here we focus on pairs of configurations having a finite number of differences (diamonds) like in the celebrated Garden-of-Eden theorem. First, we show that it allows to study expansive-like phenomena where classical notions of chaotic behavior like positive expansivity or strong transitivity don't apply (in reversible CAs for instance). Second, we establish that for a large class of linear CA, diffusion of diamonds is equivalent to randomization (a large class of probability measures converge weakly towards the uniform measure under the action of the CA). Both properties are also equivalent to the absence of gliders in the CA. Finally, we give examples of reversible linear CAs that are strong randomizers (the convergence towards the uniform measure is simple and not only in Cesaro mean). This strong behavior is however provably impossible with linear CA having commuting coefficients (e.g. linear CA over cyclic groups).

Forewords. This extended abstract is based upon unpublished works in collaboration with A. Gajardo and V. Nesme on one hand (pre-expansivity), and B. Hellouin de Menibus and V. Salo on the other hand (randomization).

What Automata Can Provide a Medium for Life?

Tommaso Toffoli[(✉)]

Boston University, Boston MA O2215, USA
tt@bu.edu

Abstract. Hadn't this question already been answered? We all know about computation-universal Turing Machines. And we know that any such machine can simulate a space-time dynamics not unlike von Neumann's cellular automaton, which is computation- and construction-universal and among other things can play host to self-replicating machines. And that self-replication sprinkled with a bit of randomness should inexorably lead to descent with variation, competition, and thence to evolution and all that.

And note that the state of the art has much advanced in the fifty years since. "So?" Enrico Fermi would have asked, "Where are they?"

It turns out that life is by its very nature a marginal, fragile, and ephemeral kind of phenomenon. For a substrate or a "culture medium" to be able to support it, computation- and construction-universality are necessary—but by no means sufficient! Most automata (including, I suspect, Conway's very game of "life") will go through their entire life course without ever originating anything like life.

What questions, then, should we ask of a prospective medium—be it a Turing machine, a cellular automaton, or some other kind of automaton—that will probe its capabilities to originate and/or sustain some form of life?

A Brief Tour of Theoretical Tile Self-Assembly

Andrew Winslow(✉)

Université Libre de Bruxelles, Brussels, Belgium
awinslow@ulb.ac.be

Abstract. The author gives a brief historical tour of theoretical tile self-assembly via chronological sequence of reports on selected topics in the field. The result is to provide context and motivation for these results and the field more broadly.

Contents

Invited Papers

Propagation, Diffusion and Randomization in Cellular Automata

Guillaume Theyssier[✉]

Insitut de Mathématiques de Marseille (CNRS, AMU, Centrale Marseille), Marseille, France
guillaume.theyssier@cnrs.fr

Abstract. A large part of the study of cellular automata dynamics consists in comparing pairs of configurations and their orbits. Here we focus on pairs of configurations having a finite number of differences (diamonds) like in the celebrated Garden-of-Eden theorem. First, we show that it allows to study expansive-like phenomena where classical notions of chaotic behavior like positive expansivity or strong transitivity don't apply (in reversible CAs for instance). Second, we establish that for a large class of linear CA, diffusion of diamonds is equivalent to randomization (a large class of probability measures converge weakly towards the uniform measure under the action of the CA). Both properties are also equivalent to the absence of gliders in the CA. Finally, we give examples of reversible linear CAs that are strong randomizers (the convergence towards the uniform measure is simple and not only in Cesaro mean). This strong behavior is however provably impossible with linear CA having commuting coefficients (e.g. linear CA over cyclic groups).

Forewords. This extended abstract is based upon unpublished works in collaboration with A. Gajardo and V. Nesme on one hand (pre-expansivity), and B. Hellouin de Menibus and V. Salo on the other hand (randomization).

1 Introduction

Cellular automata are dynamical systems and standard definitions and tools from the general theory can be considered to study them [2–4,14]. Concerning topological dynamics for instance, the classical notion of (positive) expansivity has been applied to CA giving both a rich theory in the one-dimensional case [1,4,12] and a general inexistence result in essentially any other setting [13,17]. Besides, in the ergodic dynamics settings, a lot of work was accomplished to show examples of randomizing behaviors (sometimes depicted as a kind of the second law of thermodynamics): a large class of probability measures (weakly) converge towards the uniform measure under the action of the CA [5,6,9,15,16] (see [14] for a general review of ergodic theory of CA).

Cellular automata also have a particular structure and properties that don't necessarily admit sensible formalization in the general setting of

M. Cook and T. Neary (Eds.): AUTOMATA 2016, LNCS 9664, pp. 3–9, 2016.
DOI: 10.1007/978-3-319-39300-1_1

topological/ergodic dynamics. For instance, the space of configurations allows to define the notion of diamonds: two configurations that differ only on finitely many positions of the lattice. The Garden of Eden theorem, which has a long history [2,3,7,10,11] and is emblematic of this CA specific theoretical development, then says that surjectivity is equivalent to pre-injectivity (injectivity on diamonds) if and only if the lattice is given by an amenable group.

In this extended abstract we introduce basic definitions about possible evolutions of diamonds under the action of a CA: gliders, pre-expansivity, diffusion. We then show that this point of view allows to characterize (weak) randomization combinatorially and link it to pre-expansivity in a large class of linear cellular automata. We also give reversible examples with interesting behaviors in this context: pre-expansivity without positive expansivity and strong randomization.

2   Formal Definitions

We consider spaces of configurations of the form $Q^{\mathbb{Z}^N}$ for some $N \geq 1$ and Q a finite set. Given two configurations c and d, we denote by $\Delta(c,d)$ the set of positions where they differ:

$$\Delta(c,d) = \{z \in \mathbb{Z}^N : c(z) \neq d(z)\}.$$

Two configurations c,d are *asymptotic*, denoted $c \overset{\infty}{=} d$, if they differ only in finitely many positions: $\Delta(c,d)$ is finite. When $c \overset{\infty}{=} d$ and $c \neq d$, we say it is a *diamond* and denote it by $c \diamond d$.

A cellular automaton F is an endomorphism of $Q^{\mathbb{Z}^N}$ characterized by a *local transition map* $f : Q^V \to Q$, where $V \subset \mathbb{Z}^N$ is finite (called *neighborhood of F*), and defined as follows:

$$\forall c \in Q^{\mathbb{Z}^N}, \forall z \in \mathbb{Z}^N, F(c)(z) = f(v \in V \mapsto c|_{z+v})$$

2.1   Diamonds and Their Dynamics

We are interested in the evolution of diamonds under the evolution of a given CA. More precisely, we look at the sequence:

$$\Delta_F^t(c,d) = \Delta(F^t(c), F^t(d))$$

when t grows and $c \diamond d$.

If $E \subseteq \mathbb{Z}^N$ is finite, we denote by $\mathrm{diam}(E)$ its *diameter, i.e.* the largest distance (Manhattan) between any two of its elements.

Definition 1 (Glider). *A glider for a CA F is a diamond $c \diamond d$ such that there is a bound B with $\mathrm{diam}(\Delta_F^t(c,d)) \leq B$. for all t.*

By the Garden-of-Eden (or Moore-Myhill) Theorem, any non-surjective CA admits a glider: it's not pre-injective so it admits a diamond $c \diamond d$ such that $\Delta_F^t(c,d) = \emptyset$ for all $t \geq 1$. The converse is false (some surjective CA also admit gliders). In the following we are essentially interested in surjective CA.

Definition 2 (Pre-expansivity). *F is* pre-expansive *if there exists a finite observation window $W \subseteq \mathbb{Z}^N$ such that, for any diamond $c \diamond d$, there is some time $t \geq 0$ such that $\Delta_F^t(c, d) \cap W \neq \emptyset$.*

Given a set of integers $X \subseteq \mathbb{N}$ we denote by $d_+(X)$ its (upper) density:

$$d_+(X) = \limsup_{n \to \infty} \frac{\#E \cap \{1, \ldots, n\}}{n}.$$

When considering a sequence (s_n), its density is the density of its support.

Definition 3 (Diffusivity). *F is* diffusive *if for any diamond $c \diamond d$ we have*

$$\#\Delta_F^{t_n}(c, d) \xrightarrow[n \to \infty]{} \infty$$

on an increasing sequence (t_n) of density one. F is strongly diffusive *if the above limit holds without taking a subsequence of $\mathbb{N}$, i.e. if*

$$\#\Delta_F^t(c, d) \xrightarrow[t \to \infty]{} \infty.$$

We will sometimes speak about *weak diffusivity* to emphasize that it is not strong diffusivity.

2.2   Ergodic Dynamics

A (finite) pattern is a partial configuration of finite support: $u : \mathrm{Dom}(u) \subseteq \mathbb{Z}^N \to Q$ with $\mathrm{Dom}(u)$ finite. The *cylinder* $[u]$ associated to such a pattern u is the set of configuration:

$$[u] = \left\{ c \in Q^{\mathbb{Z}^N} : \forall z \in \mathrm{Dom}(u), u(z) = c(z) \right\}$$

A Borel probability measure over $Q^{\mathbb{Z}^N}$ is completely determined by a map μ from cylinders to $[0; 1]$ such that:

1. $\mu\left(Q^{\mathbb{Z}^N}\right) = 1$
2. if $[u] = \bigcup_k [u_k]$ with $[u_k]$ pairwise disjoint then $\mu([u]) = \sum_k \mu([u_k])$

A measure is shift-invariant if $\mu(E) = \mu(\sigma_z(E))$ for any (measurable) set E and any translation $z \in \mathbb{Z}^N$.

A Bernoulli measure is a shift invariant product measure in the sense that the measure of any cylinder $[u]$ is given by the product:

$$\mu([u]) = \prod_{z \in \mathrm{Dom}(u)} \mu^*(u(z))$$

where $\left(\mu^*(q)\right)_{q \in Q}$ is the probability vector on letters of the alphabet that characterizes μ. It is non-degenerate if $0 < \mu^*(q) < 1$ for all $q \in Q$. An important particular case is the *uniform measure* μ_0: the Bernoulli measure where all $\mu^*(q)$ are equal to $\frac{1}{\#Q}$.

Given a CA F and a measure μ, the image measure $F\mu$ is the measure given by:

$$F\mu([u]) = \mu\big(F^{-1}([u])\big).$$

All surjective CA preserve the uniform measure ($F\mu_0 = \mu_0$) and conversely any CA that preserves the uniform measure is surjective [3,8].

Definition 4 (Randomization). *Given a class of measures X, we say that F is randomizing if for any $\mu \in X$ and any cylinder $[u]$*

$$\lim_{n\to\infty} F^{t_n}\mu([u]) = \mu_0([u]) = (\#Q)^{-\#\,\mathrm{Dom}(u)}$$

on an increasing sequence (t_n) of density 1.

F is strongly randomizing if the above limit holds without taking a subsequence of $\mathbb{N}$, i.e. if

$$\lim_{t\to\infty} F^t\mu([u]) = \mu_0([u]) = (\#Q)^{-\#\,\mathrm{Dom}(u)}.$$

This definition depends of course on the choice of X. In the sequel and without any explicit mention, X will always be the set of non-degenerate Bernoulli measures.

3 The Ideal World of 1D Abelian Linear Cellular Automata

This section aims at developing a rather complete analysis of diamond dynamics under two strong assumptions:

- in 1D, any incoming information arrives either from the left or from the right, and must cross any large enough window;
- for linear CA, it is sufficient to study finite configurations (*i.e.* almost everywhere equal to state 0) to actually understand all possible diamonds.

Definition 5. *Let $(Q, \oplus)$ be a finite Abelian group and denote by $\overline{\oplus}$ the component-wise extension of $\oplus$ to $Q^{\mathbb{Z}^N}$. A CA F over $Q^{\mathbb{Z}^N}$ is linear for $(Q, \oplus)$ if*

$$\forall c, d \in Q^{\mathbb{Z}^N} : F(c\,\overline{\oplus}\,d) = F(c)\,\overline{\oplus}\,F(d)$$

Equivalently, such a CA can be written as:

$$F(c)_z = \sum_{i\in V} h_i(x_{z+i})$$

where $V \subseteq \mathbb{Z}^N$ is finite and the h_i are homomorphisms of the group $(Q, \oplus)$. A CA which is linear for some Abelian group is called Abelian CA in the following.

The following lemma justifies definitions of previous section, in particular the interest of restricting to subsequences of time steps of density 1.

Lemma 1. *An Abelian CA is diffusive if and only if it has no glider. In particular, a pre-expansive Abelian CA is always diffusive.*

3.1   A Characterization of Randomizing CA

A *character* of a group $\mathcal{G}$ is a continuous group homomorphism $\mathcal{G} \to \mathbb{T}^1$, where $\mathbb{T}^1$ is the unit circle group (under multiplication). Denote $\hat{\mathcal{G}}$ the group of characters of $\mathcal{G}$. Given an Abelian group $(Q, \oplus)$, the set of configurations $Q^{\mathbb{Z}^N}$ forms an Abelian group by component-wise application of $\oplus$. When considering a linear CA F for the group $(Q, \oplus)$ and a character $\chi \in Q^{\hat{\mathbb{Z}}^N}$, then $\chi \circ F$ is also a character because F is an homomorphism of $Q^{\mathbb{Z}^N}$.

$Q^{\hat{\mathbb{Z}}^N}$ is in bijective correspondence with the sequences of $(\hat{Q})^{\mathbb{Z}^N}$ whose elements are all $\mathbf{1}$ (constant map equal to 1) except for a finite number. That is, $\chi \in Q^{\hat{\mathbb{Z}}^N}$ can be written as $\chi(x) = \prod_{k \in \mathbb{Z}^N} \chi_k(x_k)$ where all but finitely many elements χ_k are equal to $\mathbf{1}$. For a given χ, the number of nontrivial χ_k in the sequence is its *rank* denoted $\text{rank}(\chi)$.

The *Fourier coefficient* of a measure μ associated to a character χ is:

$$\hat{\mu}[\chi] = \int_{Q^{\mathbb{Z}^N}} \chi d\mu = \sum_{u \in Q^D} \mu([u]) \prod_{k \in D} \chi_k(u_k)$$

where D is any finite set that contains all χ_k which are not $\mathbf{1}$. We this definition we have

$$(\hat{F}\mu)[\chi] = \hat{\mu}[\chi \circ F].$$

Fourier coefficients characterize measures in the following sense:

$$\forall u, \ \mu_n([u]) \to \mu_\infty([u]) \text{ if and only if } \forall \chi, \ \hat{\mu}_n[\chi] \to \hat{\mu}_\infty[\chi].$$

In [15,16], the class of harmonically mixing measures is considered: μ is harmonically mixing if its Fourier coefficients uniformly go to zero when the rank of characters increase, formally: for all $\varepsilon > 0$, there exists a $R > 0$ such that $\text{rank}(\chi) > R \Rightarrow \hat{\mu}[\chi] < \varepsilon$. This class contains many natural measures, in particular any non-degenerate Bernoulli and Markov measure [15,16]. The interest of the definition is that if one wants to show that a linear CA F (strongly) randomizes all harmonically mixing measures, then it is sufficient to show that

$$\text{rank}(\chi \circ F^t) \to_t \infty$$

for all character χ (or that the limit holds on a subsequence of time steps of density 1 for the case of weak randomization).

This property (weak version) was introduced in [15,16] using the adjective "diffusive". To avoid confusion, we will use the term "chi-diffusive" here.

The following Theorem gives a characterization of randomization for a large class of Abelian CA. It shows in particular that the sufficient condition of chi-diffusivity introduced is actually necessary as soon as we ask for randomization of a class of measures that contain non-degenerate Bernoulli measures. The key idea behind the theorem is that composing a character with iterations of a linear CA F over $Q^{\mathbb{Z}^N}$ can be seen as the action of a new cellular automaton $\hat{F}$ (its dual) acting on the space $(\hat{Q})^{\mathbb{Z}^N}$. It is then possible to show that F and $\hat{F}$ are close enough dynamically so that they share properties like diffusion.

Theorem 1. *Let F be a linear CA over the group $\mathbb{Z}_{p^l}^k$ for some prime p and $l, k \geq 1$. Then the following are equivalent:*

- *F has no glider;*
- *F is diffusive;*
- *F is chi-diffusive;*
- *F randomizes any non-degenerate Bernoulli measure;*
- *F randomizes any harmonically mixing measure.*

Corollary 1. *For linear CA F, G verifying the hypotheses of the previous theorem we have:*

- *if F is pre-expansive then F randomizes all harmonically mixing measures;*
- *F and G randomize harmonically mixing measures if and only if $F \times G$ does.*

3.2 Interesting Reversible Examples

In this subsection, we consider examples of Abelian CA over groups of the form $Q = \mathbb{Z}_p^2$. To facilitate notations, we represents elements of Q as vectors and use matrix notation to represent homomorphisms of Q.

Proposition 1. *Let F_3 be the linear CA over $\mathbb{Z}_3^2$ defined by:*

$$F_3(c)_z = \begin{pmatrix} 1 & 0 \\ 0 & 0 \end{pmatrix} \cdot c_{z-1} + \begin{pmatrix} 1 & 1 \\ 1 & 0 \end{pmatrix} \cdot c_z + \begin{pmatrix} 1 & 0 \\ 0 & 0 \end{pmatrix} \cdot c_{z+1}$$

where the addition is done in $\mathbb{Z}_3^2$. F_3 is pre-expansive but not positively expansive because reversible.

Proposition 2. *Let G_2 be the linear CA over $\mathbb{Z}_2^2$ defined by:*

$$G_2(c)_z = \begin{pmatrix} 1 & 1 \\ 1 & 0 \end{pmatrix} \cdot c_z + \begin{pmatrix} 1 & 0 \\ 0 & 0 \end{pmatrix} \cdot c_{z+1}$$

where the addition is done in $\mathbb{Z}_2^2$. G_2 is strongly randomizing.

We don't have any characterization of strong randomization. However, there is a natural class of linear CA where strong randomization is impossible whereas weak randomization can happen.

Definition 6 (super-Abelian CA). *A linear CA F over an Abelian group of the form*

$$F(c)_z = \sum_{i \in V} h_i(x_{z+i})$$

is super-Abelian if the homomorphisms h_i commute pairwise.

The class of super-Abelian CA includes all linear CA over cyclic groups, but also all linear CA whose coefficients are just scalars (called LCA in [6,15,16]).

The intuition in the following proposition is that commutation in a super-Abelian CA induces too many cancellations of coefficients at specific time steps, so that infinitely many iterations go back close to the initial (non-uniform) measure.

Proposition 3. *No super-Abelian CA is strongly randomizing.*

References

1. Blanchard, F., Maass, A.: Dynamical properties of expansive one-sided cellular automata. Isr. J. Math. **99**, 149–174 (1997)
2. Ceccherini-Silberstein, T., Coornaert, M.: Cellular Automata and Groups. Springer, Heidelberg (2010)
3. Hedlund, G.A.: Endomorphisms and automorphisms of the shift dynamical system. Math. Syst. Theor. **3**, 320–375 (1969)
4. Kůrka, P.: Topological and symbolic dynamics. Société Mathématique de France (2003)
5. Lind, D.A.: Applications of ergodic theory and sofic systems to cellular automata. Physica D Nonlinear Phenom. **10**(1–2), 36–44 (1984)
6. Maass, A., Martínez, S., Pivato, M., Yassawi, R.: Asymptotic randomization of subgroup shifts by linear cellular automata. Ergodic Theor. Dyn. Syst. **26**, 1203–1224 (2006)
7. Machi, A., Mignosi, F.: Garden of eden configurations for cellular automata on Cayley graphs of groups. SIAM J. Discrete Math. **6**(1), 44 (1993)
8. Maruoka, A., Kimura, M.: Condition for injectivity of global maps for tessellation automata. Inf. Control **32**, 158–162 (1976)
9. Miyamoto, M.: An equilibrium state for a one-dimensional life game. J. Math. Kyoto Univ. **19**(3), 525–540 (1979)
10. Moore, E.F.: Machine models of self-reproduction. In: Burks, A.W. (ed.) Essays on Cellular Automata, pp. 187–203. University of Illinois Press (1970)
11. Myhill, J.: The converse of moore's garden-of-eden theorem. In: Proceedings of the American Mathematical Society, vol. 14, pp. 658–686. American Mathematical Society (1963)
12. Nasu, M.: Nondegenerate q-biresolving textile systems and expansive cellular automata of onesided full shifts. Trans. Am. Math. Soc. **358**, 871–891 (2006)
13. Pivato, M.: Positive expansiveness versus network dimension in symbolic dynamical systems. Theor. Comput. Sci. **412**(30), 3838–3855 (2011)
14. Pivato, M.: The ergodic theory of cellular automata. Int. J. Gen. Syst. **41**(6), 583–594 (2012)
15. Pivato, M., Yassawi, R.: Limit measures for affine cellular automata. Ergodic Theor. Dyn. Syst. **22**, 1269–1287 (2002)
16. Pivato, M., Yassawi, R.: Limit measures for affine cellular automata ii. Ergodic Theor. Dyn. Syst. **24**, 1961–1980 (2004)
17. Shereshevsky, M.A.: Expansiveness, entropy and polynomial growth for groups acting on subshifts by automorphisms. Indagationes Math. **4**(2), 203–210 (1993)

What Automata Can Provide a Medium for Life?

Tommaso Toffoli[(✉)]

Boston University, Boston, MA O2215, USA
tt@bu.edu

Abstract. Hadn't this question already been answered? We all know about computation-universal Turing Machines. And we know that any such machine can simulate a space-time dynamics not unlike von Neumann's cellular automaton, which is computation- and construction-universal and among other things can play host to self-replicating machines. And that self-replication sprinkled with a bit of randomness should inexorably lead to descent with variation, competition, and thence to evolution and all that.

And note that the state of the art has much advanced in the fifty years since. "So?" Enrico Fermi would have asked, "Where are they?"

It turns out that life is by its very nature a marginal, fragile, and ephemeral kind of phenomenon. For a substrate or a "culture medium" to be able to support it, computation- and construction-universality are necessary—but by no means sufficient! Most automata (including, I suspect, Conway's very game of Life) will go through their entire life course without ever originating anything like life.

What questions, then, should we ask of a prospective medium—be it a Turing machine, a cellular automaton, or some other kind of automaton—that will probe its capabilities to originate and/or sustain some form of life?

1 Introduction

In this paper I dwell on certain concerns that in my opinion haven't been voiced loudly enough—if and when they were articulated at all. These concerns have to do with the *spontaneous emergence* and the *continued sustenance* of something like *life*. More specifically, if *lifelike* behavior is desired to emerge out of an *automaton-like* medium—for definiteness, a distributed dynamical system such as a Turing machine or a cellular automaton—what conditions may make this emergence and its historical persistence more (or less) likely? To wit, *what kind of dynamical laws* and *what kind of initial conditions*?

A first-year student of theoretical computer science will of course be tempted to answer my question by means of a simple, though grandiose, reductionistic plan. "What's the problem?" she may ask. And explain:

© IFIP International Federation for Information Processing 2016
Published by Springer International Publishing Switzerland 2016. All Rights Reserved
M. Cook and T. Neary (Eds.): AUTOMATA 2016, LNCS 9664, pp. 10–25, 2016.
DOI: 10.1007/978-3-319-39300-1_2

1. We have an uninterrupted four-billion-year historical record of the emergence of Life-as-we-know-it here on Earth, an average planet of an average sun of an average galaxy. The latter is one of the few dozen galaxies that make up the Local Group galaxy cluster, which in turn is an average component of the Laniakea Supercluster. The observable universe comprises ten million odd such superclusters (en.wikipedia.org/wiki/Supercluster). Thus Earth is an undistinguished place; you're likely to find one like it in the vicinity of no matter where you look in the universe.

2. We know physics—the dynamics of our universe—at a level of detail sufficient to account for the properties of ordinary matter, run-of-the-mill cosmology, the laws of geology and chemistry, and biological structures and processes.

3. We have a reasonable idea of what the state of our universe could have been like, say, ten billion years ago, well before our solar system (and thus Sun, Earth, earthly life, and all that) had emerged.

4. Finally, I just learned that we can construct a *Universal Turing Machine* (UTM). To make a long story short, according to the Church–Turing thesis, any ordinary computer, including the minuscule Intel 4004—the first microprocessor, with only 2000 transistors (today we have microprocessors with 4 billion transistors!) are *Universal Computers*. That is, they can be programmed so that, given enough memory and time, they can exactly emulate (though perhaps more slowly and less efficiently) anything that can be computed by any other physical computer.

"Given all of that," our student will continue, "I would just program my computer to run a *simulation model* of our universe. Points 2 and 3 above will give me all the laws and data I need to write my program. Even if from my simulation I don't get exactly the kind of life that we have here on Earth today, I should get something in that ballpark, since by Point 1 an environment like Earth is so common. At worst, to find life I would have to look at several places and/or try several times! Thus the procedures and parameters of my computer program, which embody Points 2 and 3 of our physical world, do specify *for an automaton* dynamical laws and initial conditions of a kind that make the emergence of life in it likely, as per your request."

If we accept the Church–Turing thesis (which I do), my student's answer will be formally correct, but only insofar as tautological—and thus essentially vacuous. In effect she would be saying, "If our universe produced life [which I'll grant], and if my simulation by means of an automaton-like medium is functionally equivalent to the running of that physical universe itself as per Points 2 and 3 [which I'll grant as well], I don't see how this automaton would fail to often come up with life."

"Fine," I would say, "If we ignore for a moment that yours is just a thought experiment—your simulation would need astronomical resources to be carried out, including maybe trillions of "universe times"—then what you are saying is correct. *Unless . . .*" "Unless what?" "Unless the emergence of life is so fiendishly dependent on the laws of physics being *just so* and the initial conditions being *just so*, that any small deviation from that would risk making us miss the sweet spot—Turing and Church notwithstanding!"

In sum, to discover (or learn) what kind of dynamical laws and initial conditions are favorable to life we need an approach that is not only more *practical* but also more *robust*. We may eventually have to accept that life depends in an *essential way* on the fine-tuning of a certain parameter, but it would be perverse and wasteful—and, I'd say, non-scientific—to start from the outset by looking for an explanation of life that depended on the fine-tuning of multiple parameters. We'd rather study different properties that address different features of life, or produce some aspects of it in isolation. Once we've gained familiarity with one property (which may be of an ON-or-OFF kind or come with adjustable parameters) and with the effect of the constraints it imposes on a system (since studying a property *is* imposing a constraint), we can try to combine two or more of those properties and see how much interesting space is left as their intersection, if any. Echoing one of Richard Feynman's last messages, "What we cannot create we don't understand."

I had the fortune to live through and personally take part in a discovery process (concerning the computation and construction capabilities of *invertible cellular automata*), that followed by and large the above strategy; this is chronicled with copious references in [5]. At the end of the day, Norman Margolus and I could conclude, "Only a few years ago, what was known about invertible cellular automata could be summarized in a few lines—and wasn't very exciting either. Today, one can tell a [much] more interesting story."

What is remarkable is that there we'd asked a question of automata theory— "Can general computation take place in an *invertible* automata medium?"—at a time when there were conjectures and (alleged) proofs that the answer would be "No!" But we said, "Wait a moment! Physics (whether classical or quantum) seems to be able to do computation under the same constraint (of invertibility) and the additional constraint of *continuity*. What is going on? How can physics do that? How can we emulate it in automata? And what (if any) do we have to give up in return?" In the end, the answer turned out to be "Yes"—we'd been able to 'throw out the bath water' without losing the 'baby'!

2   Some Historical Notes

The challenge of creating contrivances able to imitate life and thought has been recognized since antiquity, as the myths of Daedalos, Pygmalion, and the Golem attest. The increasing technological sophistication of the 1700s and 1800s, with Vaucanson's dancing and flute-playing automata (1737), Jacquard's programmable tapestry loom (1801), Charles Babbage's differential engine (1830) and his more ambitious analytical engine (the first general-purpose computer, left incomplete, 1850), Louis Braille's writing machine (a true dot-matrix printer, 1839), and innumerable other creations, both responded to and fueled a popular fascination for ever more capable lifelike mechanical creations. At the same time, it seemed evident that higher levels of behavioral complexity could only be achieved by introducing ever more sophisticated, more precise, and fundamentally more *capable* or *powerful* mechanical parts. For instance, wind-up springs were successfully replaced, as a source of motive power, by hydraulic and pneumatic actuators, by steam engines, and eventually by electric motors.

That more complicated behavior is only made possible by more capable components has been a commonly accepted view for much of human history. This perception was philosophically formalized by Aristotle (ca. 330 BC) and (in a Christian-adapted version) by Thomas Aquinas (ca. 1270). The latter argues that stones are ... just stones—they don't do anything and don't need much of an explanation. But plants grow, reproduce, and survive. For this they obviously need a special faculty—let us call it a *vegetative* soul. Animals also grow, survive, and reproduce, but in addition they move on purpose, see, feel, react, and communicate. For this, they obviously need a higher-level faculty, which one may call a *sensitive* soul. Humans do all of the above, and also have abstract thought, articulate speech, and long-term planning. For this—you guessed it—they must be endowed with an even higher-level faculty, which one may call *rational* soul. One thus postulates a hierarchy of faculties as an "explanation" of an observed hierarchy of behavioral complexities. Neat, but vacuous. I'm sure that if one wanted to explain the faculties of angels one could always ascribe them to an ad hoc *angelic* soul (incidentally, Thomas Aquinas' moniker in the trade was "Doctor Angelicus").

Sporadic attempts were made in the past to replace this ever-ascending hierarchical ladder of "faculties" or "souls" with a single, unified, reductionistic once-and-for-all hypothesis. We shall just mention Democritus' atomic theory (400 BC), later picked up by Epicurus (ca. 290 BC) and masterly elaborated by Lucretius, in his *De Rerum Natura* ("On the make-up of things," ca. 90 BC), into a comprehensive and humane naturalistic doctrine. But, as we've seen, the Aristotelian approach, which described things in a way closer to what they superficially like from a human standpoint—and thus mentally less demanding—remained until recently the more widely circulated. Under the guise of Neo-Thomism, it was given official sanction as a philosophical framework for catholic theology (by Pope Leo XIII's encyclical *Aeterni Patris*) as late as 1879!

One could quibble whether the promotion, say, from "vegetative" to "sensitive" was a fundamentally *discrete* step, requiring the infusion of a higher-level soul, or could be achieved by a gradual, *continuous* process, but it seemed evident that the gap between "vegetative" and "inert," between living and non-living, was a major one, unbridgeable without external intervention. Indeed the distinction between "inorganic" and "organic" chemical compounds originally rested on whether they could be synthesized in the laboratory or only by living organisms. This gap was famously breached in 1828 by Friedrich Wöhler's discovery that urea—a byproduct of animal metabolism—can be produced in the lab from inorganic starting materials, thus helping put to rest the widely held doctrine of "vitalism."

As is well known, an analogous breakthrough occurred in the 30's, with independent but convergent results by Gödel, Post, Church, and Turing. Computation can be mechanized, and moreover reduced to primitive mechanical components of utter simplicity—basically, just wires and NAND gates. Granted that faculty, and thus anywhere above a very low threshold of *quality*, as it were, computability is just a matter of *quantity*—how long your program has to be to describe how to compute a certain function in a given programming language. You don't have to add new constructs to your language—or new modules to your microprocessor—to compute a more complex function, or, for that matter, *any* computable function. The *intelligence*—the "white collar"—needed for a computation can be distilled and captured once and for all onto memory as a text program. The repetitive, low-skilled *labor*—the "blue collar"—is provided by a microprocessor built once and for all and which will do for *any* task. To have the computation performed, one just has to provide as much of those raw, unstructured, "passive" resources—energy (or food), memory space (or papyrus), and time—as required.

Think of the swarm of human calculators organized into a "hive computer" by Richard Feynman during the Manhattan Project. A numerical computational task would be broken down into subtasks that individuals of modest mathematical skills could carry out with pencil and paper—some sort of glorified accounting (have you ever seen a 'calculus pad'?). To perform their job, these human calculators didn't even have to know what the overall product would turn out to be. This was actually a *design feature* of the whole outfit—you cannot accidentally let out a secret if you were not told more than you strictly need to know!

Just as Turing & Co. had "mechanized thought" in the 30's, in the late 40's John von Neumann tackled the problem of "mechanizing life." He had many talents as a mathematician, physicist, and systems scientist. The precarious alternation of success and tragedy for Jews in Central Europe must have imprinted him with a frantic urgency not only to seek personal security as a "court magician" (his productivity as an indispensable scientific consultant for government and enterprise was immense), but also to investigate in a systems-theoretic way the problems of self-preservation and self-reproduction.

So from models of abstract thought he turned for a brief time to models of material life—movement, action, construction, growth. self-reproduction. Steering him away from models based on differential equations and moving mechanical parts, his colleague Stan Ulam convinced him of the practicality of using instead automata-like models of space, time, and dynamics. So were born *cellular automata*—a new incarnation of discrete recurrence relations—within which he sketched a strategy for designing self-replicating spacetime structures. As soon as such a strategy appeared viable (see Sect. 3.1), he turned to different projects. His orphaned project remained in the care of his assistant (the scholar Arthur Burks) who eventually completed it and wrote an extensive reasoned report about it [9].

It is a symptom of von Neumann's urgency that, in spite of his being a towering figure in theoretical physics, his cellular automaton substrate is used for an empirical—macroscopic and phenomenological—model. He tried to achieve his goal, self-replicating structures, with the "greatest economy"—that is, the "cheapest and dirtiest"—of means. He put very little theoretical physics in his cellular automata—just local interactions in a space and time framework. No energy, no inertia, no action and reaction, no invertibility, no thermodynamics or second law. In this sense, cellular automata were not only left as orphans, but also deprived of much of the rich inheritance they could have expected from such a father. It took another forty to seventy years for this neglect to be remedied [3–5].

I'm not trying to be a curmudgeon. Von Neumann *created from scratch* something that shows *some* of the essential aspects of life (we'll have to say more about this). Our problem, however, is different ("What laws and initial conditions are likely to lead to the *spontaneous emergence* of life within an automaton?") and presumably much harder. Its solution would be proportionately more rewarding. It might even provide useful suggestions for the preservation as well as the expansion of our Earth's life experiment.

3   Methodological Problems

Here I should discuss a number of methodological problems, illustrating each by one or two examples.

3.1   von Neumann's Fixed-Point Trick

I will start by giving credit to von Neumann on a point where most credit is due to him.

An important aspect of von Neumann's strategy is that his creatures happen to be *self-replicating* even though they are not literally *self-copying*. It is to his credit that, as a systems scientist, he quickly homed in onto one of the most robust and dependable strategies for self-reproduction, anticipating the discovery of ribosomal function by Zamenicnik [12] and the structure of DNA [11] by Watson and Crick [11]. The famous last sentence of [11], "It has not escaped our notice that the specific pairing we have postulated immediately suggests a possible copying mechanism for the genetic material," was a noncommittal way of stating a conjecture that involved the entire cell-reproduction scheme.

In a nutshell, to reproduce you should not try to copy your *phenotype* (your body); that is hard, impractical, and unreliable. You rely instead on your phenotype to already contain a *genotype* (a recipe or set of instructions) for building *some* phenotype (not necessarily like yours) and equipping it with a copy of the original recipe itself as a genotype. Besides containing this recipe, your genotype should be able to read it and carry it out, so as to construct a new complete phenotype. If this "daughter" phenotype itself happens to be able to read and execute its recipe, then its own daughter will happen to be identical to a copy of itself—even though this daughter was not *copied* but *made from a recipe*—and so forth, generation after generation.

With this process, you may not have ended up reproducing yourself (though after the second generation your descendants *will* reproduce *themselves*). However, if the genotype carried by phenotype X *happened* to be a recipe for the very phenotype X that carried it, then self-reproduction will start with the first generation. In the latter case, the phenotype cum genotype happens to be a *fixed point* of a *recurrence relation* (the chain of generations).

Our problem with von Neumann's brilliant insight is merely that it is not very relevant to the *first stage* of the emergence of life—though it becomes extremely relevant at later stages. (In fact, one may wonder whether *any* later stages of evolution would gamble with doing without it.)

To come to the point, observe that von Neumann's self-replication strategy requires a *matched combination* of an active-fabricator phenotype and a passive-recipe genotype. For the strategy to be successful, that genotype must *happen* to be just what is needed to direct the fabrication of that very phenotype. This requirement is clearly a tall order, as it requires a lot of pre-ordained coordination, and thus is not likely to have been the reproduction mechanism used right at the origin of Earth's life (or of any other emergent life). This mechanism must have been derived later, by a succession of cautious "make before break" steps (cf. [2]).

3.2   Computation Universality Is Not Enough

Computation universality of a medium is essential for it to support life, but that does not mean that just proving the medium's universality by embedding a Universal Turing Machine (UTM) into it makes it particularly likely to support the spontaneous emergence of life in such a way that I can observe and study it.

The argument, already sketched in the Introduction, is simple. It is true that if you put a UTM in my computing medium and show me how its tape is realized within this medium, and similarly for the machine's head, and explain to me how the machine's states are encoded in the head and the tape symbols on the tape, and finally show me the head's transition table (its "microcode"), then I'll be able to program this machine. With an appropriate programming and data-entering effort, I should be able to instruct this machine to simulate, say, a 7-dimensional cellular automaton, a neural network, a chess-player self-learning machine, or, to any required approximation, a continuous mechanism like the solar system. In principle I could even program the UTM to simulate *our entire universe* to study the emergence of life in it, perhaps with a slow-down factor of only 10^{30} and on a scale of 1,000,000,000:1 (so that I'll have to borrow another universe with a volume 10^{27} times as large as mine just to provide the machine's tape, and I would have to hibernate myself and wake up every few billion years just to see one second's worth of simulation progress. But after only a few of those seconds my universe, the UTM in it, and myself will have already vanished in a puff of smoke!

Of course if my computing medium is a high-definition 3-D cellular automaton that can run a Giga-step/s without requiring the power of a huge hydro-electric plant, and if I can easily manipulate its state and visualize its contents, then I might be able to run on it certain focused experiment that I could not easily perform on a conventional computer. But, at that point, I would be *very* surprised if its designers hadn't thought of designing computation universality into its mechanisms at the outset—at virtually no extra cost.

Computation universality is so cheap that it would be a crime not to have it everywhere, but its real worth is as a conceptual tool—it doesn't magically turn a circuit of a dozen gates traveling up and down a tape into an hypercomputer.

3.3 The Fragility of "All Mine!"

Let's consider a luxuriant millenary oak forest. It is conservatively estimated that the average oak will have produced in its lifespan 10 million acorns. How many of these (averaging over all trees) will grow into a new oak? [Think by yourself for a moment; the answer is given a couple of paragraphs down].

Lisp in its pure form is an autistic's dream. It neither produces nor feels any side-effects. A pure Lisp program is a static hierarchical tree of text and its execution leads to the growing in real-time, by iteration and recursion, of a dynamic execution tree of procedure calls and data. The Lisp program is a "Master of the World." It can branch out and out indefinitely, reaching out and spreading at will into an infinite empty space that is ready to be occupied and colonized. There are no competitors, no partners, no superiors. No noise, no disturbances, no unforeseen events, nothing beyond one's control. No alien footprints on the beach!

The essence of life, on the other hand, is competition—not in the sense of the Nietzschean superman, but in the humbleness of presuming approximate *symmetry* with our competitors (the "Copernican principle"), in the recognition that for everyone to be above average is a contradiction in terms. In a world of all heroes, no one is a hero! In life, success in competition mostly entails being adept at sharing, collaborating, and compromising. The reason for all this is very simple—we live in a polynomial spacetime, not an exponential one! In our world, the only long-term exponential growth is that with base one: $1^n \equiv 1$.

To come back to the oak-and-acorns puzzle. This is a millenary forest, so a climax forest—one that has attained dynamic equilibrium. For any acorn to grow into an oak, one oak must have died (lightning, disease, old age) to leave room for it. The lavish investment of 10,000,000:1 is to insure that "the moment one parking spot gets freed, one of your proxies will be right there to take it before somebody else does."

The above considerations are aptly illustrated by the three examples below. Scenarios that were purportedly designed to be "life-ready"—and for all I know they still might be—but within which "life" was mostly sought by means of a Nietzchean approach. The latter, as we hinted above, was bound to founder.

von Neumann's self-replicators. In von Neumann's cellular automaton, an initial self-replicator will construct a second one next to it; after this, the first will go dormant while the second will build a third one, and so on, eventually resulting in a whole row of frozen structures except for an active "tip." What will happen if we plant in this landscape a second "seed" at some distance, oriented perpendicularly to the first? When the tip of this new row impinges on the first row, a collision will occur whose consequences will depend on the relative space and time phase of the two rows and on the actual rules that are in force (von Neumann never completed his project, and many of his followers devised different variant completions [10]). There are of course in von Neumann's project no provisions for his self-replicating automata to deal with obstacles, to fight with one another, to repair themselves, to reuse scattered materials. Depending on a balance between the construction and destruction powers given to different parts of the machinery, we may imagine that the typical resolution will be a clogged up jam, a structure that shrinks until it disappears, or a runaway structure that spreads uncontrollably.

Of course one can resign oneself to using a single row of these self-replicating structures and adapt it to behave like a Turing Machine, in which case one has total control over it *as a Turing Machine*, and we know one can simulate a whole universe with it, just as we saw in the Introduction, and yet, for the same reasons discussed there, be nowhere nearer to getting an insight into the emergence of life—in spite of the self-replicating capabilities of the elements that make up the substrate of this Turing machine. But what is the point, then? The realizations described in [10] mention *billions* of time-steps for a single replication cycle! Whatever computing I need to do, I'd rather use a general-purpose computer by default, and turn to a dedicated processor if I indeed need to simulate a special architecture like a neural network or a cellular automaton.

To effectively use a von Neumann-like cellular automaton as a tool for research on the emergence of life, one will first have to plan (or let evolve) a whole consistent *ecology* that can take advantage of that architecture (even if only at the level of mental experiment). Otherwise all one will achieve is the typical outcome of impatient kids playing with a "chemistry kit." Stinks, explosions, broken glass, the more interesting chemical run out first, and eventually mom will come, say "Why don't you go and play outside?" and proceed to clean up the mess.

Conway's Game of Life. Much more attention to developing an empirical experimental playground was given in Conway's game of Life. It has simpler rules, deliberately tuned somewhere between too many and too few "births" and similarly between too many and too few "deaths." It has an interesting near-equilibrium ecological statistics, with species like semaphores, blinkers, pinwheels, gliders, glider-guns, etc. This game developed a veritable world-wide cult, and so we are in possession of a large number of interesting constructions and observations. It is computation-universal, in the sense that one can assign to it an initial configuration that acts like a Universal Turing Machine (see mathworld.wolfram.com/GameofLife.html for details). What more shall we demand?

As long as we choose to use the game of Life as a Turing machine, see my response to the similar proposal for von Neumann's cellular automaton and that given in the Introduction. Universality notwithstanding, what a waste to use an entire two-dimensional parallel-processing universe for building a sparse and slow one-dimensional UTM where only one spot at a time is active (i.e., where the head is "now"). Building a UTM with Life was of course a clever tour-de-force as a way of proving the host environment's universality. But after that one should move on.

And here is where Fermi's question, "So where are they?" is appropriate. Tens of thousands of people must have striven to get some form of "life" emerge out of Life, and so far have failed. This is certainly circumstantial evidence that Conway's idea was stimulating—but the game of Life should not be regarded as a sacred relic to be held under a glass bell.

As for using the game of Life as an ecological universe for exploring the emergence of life, my impression is that it gets gummed up much too soon, as you will convince yourself if you look at a square of say, $10,000 \times 10,000$ cells started from a random initial condition and run it for 100,000,000 steps. Conway's Life is too jumpy on short-range interactions, and too refractory to long-distance interactions. One reason for that is that the non-invertibility of Life—the fact that it has orbits that *merge* (see [7] for a more thorough discussion), leads to a gradual "draining" of the effective space state, until the rule becomes effectively invertible and the orbits that are left are mostly short closed ones.

So we should not stop at that particular game of Life, but follow Conway's *spirit* and develop versions that have a well-argued promise for that "equilibrium near the edge of chaos" that life seems to thrive on. This rationale is explained in Sect. 4.

4   Specific Ergodicity

Possibly the parameter that most directly affects the capabilities of a distributed dynamical medium, like a cellular automaton, to support the emergence of complex structures, is the dependence of interaction strength on distance.

Complex, coordinated, knowledge-rich structures require memory—lots of it. And the characteristic property of memory is, of course, resistance to disturbances: I want to find in it what I put in it in the first place! Another important property of memory is, of course, ease of access, primarily reading access, but secondarily also writing access. In our DNA, which is vitally important, writing is done only at fabrication time; thereafter DNA is read-only. Moreover, DNA never leaves the tight shelves of the nucleus where it is stored. For work memory, the librarian will only allow you to come in person and copy a few pages at a time; you are free to use and reuse that, and throw it away when you're done. On the other hand, neural memory, which needs continual real-time upgrading, is stored at synapses, where it can be modified, though only gradually. Moreover, the same neuron firing is registered—even though with different interpretations—by thousand of synapses; so there is some redundancy there.

Other parts of an organism have to be constantly alert and active—a continual state of receptivity and change. But still you want to be selective about *what* is actually allowed to impinge on you, and your movements must be checked by feedback, lest you break everything you touch or you are broken by everything that touches you!

Organic life has managed to synthesize different materials wonderfully attuned for different functions—bone, skin, hair, mucus, hydrochloric acid, fat, crystalline lens, hemoglobin—and able to appropriately modulate and filter internal and external solicitations. On the other hand, a single general tuning knob ranging from REFRACTORY to HYPERACTIVE for the universe as a whole would be convenient for discovering the most effective MID-RANGE tuning. We have explored such a parameter, called *specific ergodicity*, in [6].

Basically, conventional ergodicity is a YES/NO parameter (a system is *ergodic* if all its states belong to a single orbit, or are in some probabilistic sense accessible from any other point). On the other hand, *specific ergodicity* is a parameter $0 \leq \eta \leq 1$, for distributed systems like cellular automata for which it makes sense to speak of an arbitrarily large wrapped-around patch governed by the same local rule, and to take the limit for the size of the patch going to infinity. In other words, for a situation where one can speak of a distributed system as a *material* rather than an *object*.

An invertible cellular automaton has $\eta = 0$ if every state has its own orbit of length 1, so that there is no interaction whatsoever between sites—every site is an isolated "monad"—and $\eta = 1$ if all states are on a single orbit (or a small number of orbits of almost maximal length), and thus in the long term the state of any site will be correlated with that of any other state—you can never be free of disturbances by neighbors no matter how distant.

A sample of well-known cellular automata gave values for η scattered over the whole range $[0, 1]$; thus, whatever this parameter means, the parameter appears to be *informative*—to be able to tell classes of systems apart. (If most of a library's books had a call number that began with P, the first letter of a book's call number would carry very little information—and so be of little use in classification.)

5 Emergence

Emergence is a term that scientists use in a specialized sense: very briefly, a pattern is *emergent* if it spontaneously arises (literally, "comes to the surface") from an underlying patternless substrate. Emergence is a class of statistical phenomena associated with a dynamical system.

A canonical example is the regular pattern of dunes that naturally develops out of a flat expanse of sand under the action of a steady breeze. Why do we get *ripples* if both sand and wind are *smooth* to begin with? And who insured the ripples' uniform spacing and specified its pitch? Based on animate beings' practices they were familiar with, the ancients fancied that the sun was pulled across the sky by a cosmic charioteer and lightning bolts were supplied to Zeus by a cosmic forger. Just as well, they might have fancied that dunes are raised by a cosmic potter tracing grooves with his fingers, or are turned up by the plowing of a cosmic farmer—techniques called respectively "fluting" and "furrowing."

But in fact, given the wind, the dunes arise by themselves with no external help, planning, or coordination. What happens if we come with a tractor and flatten the whole dune field? This has been done, and in a few days the dune field re-grew, with the same orientation and spacing—even though not necessarily with the same *phase*. (For example, the new ridges may have been arisen where in the previous pattern there occurred valleys instead.)

The most trivial form of emergence is the approach to thermal equilibrium in a homogeneous medium—the very phenomenon that led Fourier to invent his transform. When two blocks of copper at different temperatures are brought into intimate contact, the initially sharp step-function temperature profile along the resulting bar immediately starts smoothing into a sigmoid, which keeps stretching and flattening, until, in the limit of equilibrium, the temperature profile reduced to a constant function—a horizontal straight line. In Fourier's analysis, the short-wavelength components of the temperature plot decay fastest, until "every valley shall be filled, and every mountain ... brought low," at all scales.

What's important is that this kind of macroscopic behavior happens all by itself—it need not be programmed in microscopic detail. This behavior is *robust*; even under perturbations, valleys will still be filled and mountains levelled. Molecules do not have to coordinate with one another. In fact, Fourier's analysis shows that they'd better not, because this particular emergent behavior is symptomatic of *linearity*, and thus of *noninteraction* between modes.

By its nature, the emergent phenomenon of thermal diffusion is *transient*—a one-act show. However, if we put a copper bar into contact at the two ends with two heat reservoirs at different temperatures, so that heat will be continually "pumped" into one end and "drained" from the other, then the temperature profile will converge to a *sloping* straight line. If we change the temperature difference, the slope will change. If we chill the block by splashing water on it, the temperature profile will be altered, but it will automatically revert to a sloping straight line soon after all the water has evaporated. Our temperature profile behaves like a stretched rubber band in a tub of molasses. You can pull it up and down and sideways, but when let go it will recover its equilibrium profile. In this case, the emergent state is *dynamically* maintained thanks to the flow of energy set up by the pump. Again, no micromanagement is needed for all of that, no exotic compounds or precisely machined part, no delicate assembly. It all just happens by itself, dependably resiliently, and predictably. If we may have a complaint, it is that what happens is only indirectly determined by us. If we get a sigmoid, we can't ask, "How about a sinusoid, instead? Emergence is more like a Chinese diner, where we can only order by number, than an *à la carte* Parisian restaurant. (Incidentally, Galileo was convinced that the shape of a rope hanging between two points was a *parabola*, which he was most familiar with; now we know it's a *catenoid*, very similar but not quite the same. It was not up to him to choose.)

Take now the flow of water through a pipe, or of a river between its banks. In either case the flow is maintained by a pressure difference, actively maintained by a pump in one case and the riverbed slope in the other.

As long as the water flows slowly, the velocity profile across the channel will be a smooth parabola, from a maximum at the center to zero at the edges; another nice mathematical shape spontaneously emerging from a compromise between friction between adjacent layers of water and friction between water and the edges of the channel—not a shape deliberately drawn by a cosmic draftsman.

Things get exciting when the pressure is increased and the flow runs faster. The layer of water running close to the bank develops lateral instabilities—it starts undulating. As the water speed increases, these undulations break up into vortices. Soon larger vortices develop smaller satellite vortices, and so forth, so that if you observe carefully you'll see a whole fractal hierarchy of them, spanning the range from macroscopic to microscopic. These vortices have an identity and a life of their own: they form and melt away, grow and shrink, they collide and annihilate one another, they "calve" daughter vortices, and so forth. They also may have "memory" and hysteresis: as you increase the speed, it may take a while for vortices to develop, but once they are there they will linger on even after you've reduced the speed to quite below what it was when they were born.

A similar phenomenon occurs with a dripping faucet, or when you overblow a note in a recorder. Also, the breakup of a thin stream of water into droplets may be encouraged by external vibrations, so that one can obtain in that way a very sensitive detector of ground temblors. What's more, as you adjust a faucet's flow, a train of regularly spaced droplets may break up into a train of paired droplet, from plic, plic, plic to plic plic, plic plic, plic plic; and this doubling may reduplicate, so that you may get a train like $((..)(..))((..)(..))$. In some sense, this system has "learned" to count.

Here we've barely scratched the surface of emergence. It's an extremely rich and surprising world. On one hand, the ecologic "bestiary" supported by one substrate—say, the blinkers, gliders, and pinwheels of the game of Life—may be totally alien to those of another substrate—say, Norman Margolus's "critters" or Charles Bennett's "scarves" [4]. On the other hand, emergent phenomena are statistical phenomena, and naturally fall into a hierarchy that is basically one of combinatorial dynamical patterns. Diffusion, waves, cyclic rankings, frustration, annealing, predator-prey equation, hydrodynamics, single body "collisions" (like nuclear decay), two-body collisions, multi-body collisions, conservation on networks (like Kirchhoff's laws), relaxation oscillators, excitable systems, and so forth. Thus, even if the material substrate is different, system-level behaviors may obey similar laws and thus belong to the same equivalence class of phenomena.

There is an analogy here with the functions and the differential equations of physics—which, incidentally, are themselves mostly echos of emergent combinatorial phenomena. In principle, you could have all sorts of functions and differential equations, but in practice only a handful of them are truly common and ecumenical.

As we've seen, emergent behavior needs to be driven by a "pump," or what's called an "entropic cascade." (In the case of thermal equilibration between two bodies, the thermal difference only provides a single-use "battery" that ends up discharged when this difference reaches zero.) When an ordered arrangement of system A tumbles down the hill of increasing entropy turning into a more disordered arrangement A', some of A's "predictability" can be tapped and used to "pump up" another system from a more disordered state B' to a more ordered state B. An important case is when the peculiar order in A is of no special interest to us *per se*—all that we care is the *amount* of order that it contains. By coupling system A to system B we can extract some of the *generic* predictability of A and convert it into an equivalent amount of a *specific*—more desirable—form of predictability in B. (For example, if A is energetic but inedible wood that releases fire as it decays into A' (ashes and $CO2$), the fire can be used to bring a dish B' of raw meat or potatoes to a more edible state B.) As soon as the pump stops, the emergent structure "propped up" by it may of course collapse.

6   Life and Evolution

Life—and the opposite side of the coin, namely, evolution—is literally the *runaway daughter* of emergence.

Given a reliable entropic pump (this is really what is usually meant when one speaks of a "source of energy"), such as our Sun for the Earth, an emergent pattern based on it—artificial and propped up as it may be—constitutes an additional environmental niche available for habitation by materials, processes, and reactions, and may very well be one that provides tools and goods at a higher structural level than those provided by the base environment.

For example, even though the native language of a computer is its *machine* language, once an *assembler* for that machine language is available and can be run on that computer, one can program the same computer much more conveniently and productively at the assembler language level. In a similar way, C is a generic, platform agnostic, higher-level programming language. No computer runs C natively, but C *compilers* have been written to translate C to the machine language of different computers. And one can program more expressively, compactly, and productively in C.

Today. most of the Information Technology industry relies on different emergent levels of software environments, with enormous advantages in productivity, reliability, documentability, and often (though not always) in computer performance. Of course, all of today's civilization runs exclusively on the entropic pump provided by the sun either directly or via underground caches.

The runaway aspect of life mentioned above is that, starting about four billions years ago, life has managed to exploit natural emergent environments first, and then created new emergent environments within itself, originally at a sluggish rate, but then faster and faster (see Nick Lane outstanding books [1,2]).

In my opinion, the best way to study the emergence of life—beside the study of current and paleontological life on Earth—is to design environments representing versions of physics *stylized and domesticated* at different levels (we still don't know how much we can get away with) and optimized for the support of emergent environments. It is not terribly important at this point to create special-purpose hardware platforms for these environments; software platforms will do until we have a better idea of what we want.

An important issue in this context is to choose whether to stress (a) Non-invertible models of physics, such as von Neumann's, Bank's, Conway's, and Wolfram's cellular automata, whose merging of trajectories, as we've seen in Sect. 3.3 [Conway], provide a free, built-in entropy pump at least initially; (b) Models of physics which retain invertibility and provide an entropy pump through a distinguished away-from-equilibrium initial configuration (one containing the equivalent of a tame Big Bang, as it were); or (c) Models which replace an internal entropy pump by contact with an external thermal reservoir, as is routinely done to standardize thermodynamics arguments.

7   Conclusions

We all have learned a lot about modeling life within an automaton from the past two generations, but still can't get rid of Fermi's taunting question, "So, where are they?" I'm afraid that before trying to impose our wishes on an automaton we should more humbly inquire of it, "What is it that *you* would like to do?" and then try to build on that answer.

I'm reminded of the Flea Circuses that were still seen in country fairs as late as fifty years ago (have you ever seen Charles Chaplin's *Limelight*?), where an old man with a glass-topped tray-box hanging from his shoulders would show and illustrate to an audience of "children, soldiers, and servants" the acrobatics of fleas he kept in the box. The fleas clearly didn't "understand" his commands, but he somehow managed to anticipate the kind of things they'd more likely do. He knew them, he cared about them, he "understood" them. He would build his show on the flea-y things the fleas would naturally do. I'm sure he could have made a working computer out of jumping fleas, with the fleas still "thinking" that they were doing their natural flea-y things (and that's indeed the only things they could be doing) instead of being part of a computer. Just like the seals in Sea world or even the "computers" in Feynman's team. To use them well you have to know them well and treat them well.

References

1. Lane, N.: Life ascending: the ten great inventions of evolution, Norton (2010)
2. Lane, N.: The vital question: energy, evolution, and the origins of complex life, Norton (2015)
3. 't Hooft, G.: The cellular automaton interpretation of quantum mechanics, December 2015. arXiv:1405.1548v3
4. Toffoli, T., Margolus, N.: Cellular Automata Machines: A New Environment for Modeling. MIT Press (1987)
5. Toffoli, T., Margolus, N.: Invertible cellular automata: a review. Physica D **45**, 229–253 (1990)
6. Toffoli, T., Levitin, L.: Specific ergodicity: an informative indicator for invertible computational media. In: Computing Frontiers 2005. ACM (2005)
7. Toffoli, T., Capobianco, S., Mentrasti, P.: When–and how–can a cellular automaton be rewritten as a lattice gas? Theor. Comp. Sci. **403**, 71–88 (2008)

8. von Neumann, J.: The general and logical theory of automata. In: La, J. (ed.) Cerebral Mechanisms in Behavior–The Hixon Symposium, pp. 1–31. Wiley (1951)
9. von Neumann, J.: The Theory of Self-Reproducing Automata (Arthur Burks ed.). Illinois (1966)
10. http://en.wikipedia.org/wiki/Von_Neumann_universal_constructor
11. Watson, J., Crick, F.: A structure for deoxyribose nucleic acid. Nature **171**, 737–738 (1953)
12. Zamecnik, P., Keller, E.B.: J. Biol. Chem. **209**, 337–354 (1954)

A Brief Tour of Theoretical Tile Self-Assembly

Andrew Winslow[✉]

Université Libre de Bruxelles, Brussels, Belgium
awinslow@ulb.ac.be

Abstract. The author gives a brief historical tour of theoretical tile self-assembly via chronological sequence of reports on selected topics in the field. The result is to provide context and motivation for the these results and the field more broadly.

Introduction. This tour covers only a subset of the research topics in theoretical tile self-assembly. It is intended for readers who are familiar with the basics of the field and wish to obtain a better understanding of how the multitude of models, problems, and results relate. As such, it is neither a survey nor an introduction; for these, the reader is referred to the excellent works of Doty [18], Patitz [37], Woods [50], and Winfree [47]. Moreover, it does not cover work in experimental DNA tile self-assembly.

The aTAM of Winfree (1990s). It is common for work on theoretical tile self-assembly (hereafter *tile assembly*), to begin "In his Ph.D. thesis, Winfree [47] introduced the *abstract tile assembly model (aTAM)* ...". The ubiquity of this opener matches the importance this work plays in the field: it is the point of conception, and nearly 20 years later, its reading connotes initiation to the area. Moreover, the sustained popularity of tile assembly is due in large part to the elegance and hidden intricacy of this original model. Such intricacy is perhaps most crystallized in a simple yet devious question: is universal computation possible in the aTAM at temperature 1?

As with any research, the concept of the aTAM and its experimental implementations were not in isolation. Several other models of (linear) DNA-based computation also introduced around this time, including the filtering-based models of Adleman [2] and Beaver [8] and the splicing systems of Pâun, Kari, Yokomori, and others [27,39,51].

Benchmark Problems (2000–2004). Beyond defining the aTAM, the Winfree thesis also contains a proof of the computational universality of the aTAM at temperature 2. This result established algorithmic universality, but not the ability to assemble shapes *efficiently*, i.e., using systems of few tile types. Rothemund and Winfree [41] soon established this, achieving $n \times n$ square assembly with $O(\log n)$ tile types. Following this work, the twin capabilities of universal

Published by Springer International Publishing Switzerland 2016. All Rights Reserved
M. Cook and T. Neary (Eds.): AUTOMATA 2016, LNCS 9664, pp. 26–31, 2016.
DOI: 10.1007/978-3-319-39300-1_3

computation and efficient square assembly became the de facto benchmarks for powerful models of tile assembly.[1]

Followup work by Adleman et al. [4,13] closed the small gaps in optimality left by the construction Rothemund and Winfree and introduced a new metric of efficiency: (expected) assembly time. This metric was ported from simultaneous work on the dynamics of linear assemblies by Adleman et al. [3,6] and considered in other models later [10]. Shortly after, efficient assembly of general (non-square) shapes was proved NP-complete by Adleman et al. [5], while Soloveichik and Winfree [45] established the *geometric universality* [46] of the aTAM: the construction of all shapes efficiently (if scaling is permitted).

Even beyond techniques for information encoding and construction analysis introduced in [4,45], perhaps the most persistent single contribution of work in this era was implicit conjecture in [41] that the aTAM at temperature 1 is not capable of (universal) computation, based on related conjectured lower bound of $2n - 1$ tiles to assemble $n \times n$ squares.

Error-Prone Models (2002–2011). Some of the earliest variations on the aTAM were those concerned with the design of systems robust to various errors in the assembly process. Such errors included incorrect tile attachments [11,12,42,44,49], assembly "damage" via partial deletion [44,48], temperature fluctuations [24], and unseeded growth [43]. In an ironic turn, the adversarial "seedless" growth addressed by Schulman et al. [43] was later used to achieve efficient constructions impossible with seeded growth [9]. The collection of error-prone models and results demonstrated that even small changes to aTAM yield rich new ideas. New model variations remain the largest catalyst of new work in tile assembly.

The Temperature 1 Problem (2005-ongoing). The *Temperature-1 Problem* is the most notorious open problem in tile assembly: is the temperature-1 aTAM computationally universal? The widely conjectured answer is a resounding "Obviously not!"[2], but the problem has resisted nearly all progress, permitting only negative results that use either stronger conjectures [33,34], weaker models [32,40], or the assumption of other plausible conjectures [25].

One primary difficulty is even obtaining a precise formal statement of what constitutes "computation" in tile assembly. The second is developing a proof approach that passes the "3D test": the proof must break in 3D, implied by the result of Cook et al. [14].

Computational Universality via Weak Cooperation (2007–2012). The computationally universality of the aTAM at temperature 2 and the temperature 1 problem beg the question of whether computational universality can be achieved by adding other features to the temperature-1 aTAM. Several variations of the temperature-1 aTAM were considered for which the answer proved to be "Yes". These included models using the third dimension [14],

---

[1] As discussed later, these simple challenges ultimately proved insufficient detailed for distinguishing between some powerful but unequal models.

[2] Usually accompanied by an exaggerated shoulder shrug and waving, upturned palms.

negative-strength glues [21,38], non-square tiles [26,29], and tiles with triggerable "signals" [30,36]. These models commonly exploit either "weak cooperation" in the form of repelling forces or the ability to "jump over walls".

Handedness (2008–2014). In addition to models adding features to the temperature-1 aTAM, other modifications to the aTAM were under consideration. In particular, the elimination of the seed was considered in the *hierarchical* or *two-handed (2HAM)* tile assembly models, hinted at in several settings [7,15,31] before reaching the formulation used currently [1,24]. Surprisingly, the removal of the seed causes numerous unexpected effects, including increased power [9], runaway growth [19,20], and no improvement in assembly time [10].

Intrinsic Universality (2010-present). By 2010, the introduction of new tile assembly models was occurring regularly. As previously described, many of these models obtained computational universality, but through alternative "weakly cooperative" means. With a few exceptions where direct simulation was possible (e.g., 2HAM simulation of aTAM [9]), the understanding of the relative power of these models was unsatisfyingly coarse: models are either computationally universal or not.

Adapting definitions of a geometric notion of simulation from cellular automata (see [28,35]), Doty et al. [22,23] established that the temperature-2 aTAM is *intrinsic universal* for all aTAM systems: there exists a single temperature-2 tile set that simulates the behavior of any aTAM system (when provided with a seed assembly encoding the system). Subsequent work used this new comparative metric to prove positive and negative intrinsic universality results for variations of the 2HAM, aTAM, and polygonal tile model [16,17,29,34]. In progress towards the Temperature 1 Problem, the temperature-1 aTAM was proved not intrinsically universal for the higher temperature aTAM [34], failing to match temperature 2.

References

1. Abel, Z., Benbernou, N., Damian, M., Demaine, E.D., Demaine, M.L., Flatland, R., Kominers, S.D., Schweller, R.: Shape replication through self-assembly and RNase enzymes. In: Proceedings of the ACM-SIAM Symposium on Discrete Algorithms (SODA) (2010)
2. Adleman, L.: Molecular computation of solutions to combinatorial problems. Nature **266**(5187), 1021–1024 (1994)
3. Adleman, L.: Toward a mathematical theory of self-assembly (extended abstract). Technical Report 00–722, University of Southern California (2000)
4. Adleman, L., Cheng, Q., Goel, A., Huang, M.-D.: Running time and program size for self-assembled squares. In: Proceedings of Symposium on Theory of Computing (STOC) (2001)
5. Adleman, L., Cheng, Q., Goel, A., Huang, M.-D., Kempe, D., de Espanés, P.M., Rothemund, P.W.K.: Combinatorial optimization problems in self-assembly. In: Proceedings of Symposium on Theory of Computing (STOC) (2002)

6. Adleman, L.M., Cheng, Q., Goel, A., Huang, M.-D., Wasserman, H.: Linear self-assemblies: Equilibria, entropy and convergence rates. In: Aulbach, B., Elaydi, S.N., Ladas, G. (eds.) Proceedings of Sixth International Conference on Difference Equations and Applications (2001)

7. Aggarwal, G., Cheng, Q., Goldwasser, M., Kao, M., de Espanes, P., Schweller, R.: Complexities for generalized models of self-assembly. SIAM J. Comput. **34**(6), 1493–1515 (2005)

8. Beaver, D.: A universal molecular computer. In: DNA Based Computers: Proceedings of a DIMACS Workshop, pp. 29–36. American Mathematical Society (1996)

9. Cannon, S., Demaine, E.D., Demaine, M.L., Eisenstat, S., Patitz, M.J., Schweller, R.T., Summers, S.M., Winslow, A.: Two hands are better than one (up to constant factors): Self-assembly in the 2HAM vs. aTAM. In: Proceedings of International Symposium on Theoretical Aspects of Computer Science (STACS), LIPIcs, vol. 20, pp. 172–184. Schloss Dagstuhl-Leibniz-Zentrum fuer Informatik (2013)

10. Chen, H., Doty, D.: Parallelism and time in hierarchical self-assembly. In: ACM-SIAM Symposium on Discrete Algorithms (2012)

11. Chen, H.-L., Cheng, Q., Goel, A., Huang, M.-D., de Espanés, P.M.: Invadable self-assembly: combining robustness with efficiency. In: Proceedings of the 15th Annual Symposium on Discrete Algorithms (SODA), pp. 890–899 (2004)

12. Chen, H.-L., Goel, A.: Error free self-assembly using error prone tiles. In: Ferretti, C., Mauri, G., Zandron, C. (eds.) DNA 2004. LNCS, vol. 3384, pp. 62–75. Springer, Heidelberg (2005)

13. Cheng, Q., Goel, A., Moisset, P.: Optimal self-assembly of counters at temperature two. In: Proceedings of the 1st Conference on Foundations of Nanoscience: Self-assembled Architectures and Devices (2004)

14. Cook, M., Fu, Y., Schweller, R.: Temperature 1 self-assembly: determinstic assembly in 3d and probabilistic assembly in 2d. In: ACM-SIAM Symposium on Discrete Algorithms (SODA) (2011)

15. Demaine, E.D., Demaine, M.L., Fekete, S.P., Ishaque, M., Rafalin, E., Schweller, R.T., Souvaine, D.L.: Staged self-assembly: nanomanufacture of arbitrary shapes with $O(1)$ glues. Nat. Comput. **7**(3), 347–370 (2008)

16. Demaine, E.D., Demaine, M.L., Fekete, S.P., Patitz, M.J., Schweller, R.T., Winslow, A., Woods, D.: One tile to rule them all: simulating any tile assembly system with a single universal tile. In: Esparza, J., Fraigniaud, P., Husfeldt, T., Koutsoupias, E. (eds.) ICALP 2014. LNCS, vol. 8572, pp. 368–379. Springer, Heidelberg (2014)

17. Demaine, E.D., Patitz, M.J., Rogers, T.A., Schweller, R.T., Summers, S.M., Woods, D.: The two-handed tile assembly model is not intrinsically universal. In: Fomin, F.V., Freivalds, R., Kwiatkowska, M., Peleg, D. (eds.) ICALP 2013, Part I. LNCS, vol. 7965, pp. 400–412. Springer, Heidelberg (2013)

18. Doty, D.: Theory of algorithmic self-assembly. Commun. ACM **55**(12), 78–88 (2012)

19. Doty, D.: Producibility in hierarchical self-assembly. In: Ibarra, O.H., Kari, L., Kopecki, S. (eds.) UCNC 2014. LNCS, vol. 8553, pp. 142–154. Springer, Heidelberg (2014)

20. Doty, D.: Pattern overlap implies runaway growth in hierarchical tile systems. J. Comput. Geom. **7**(2), 3–18 (2016)

21. Doty, D., Kari, L., Masson, B.: Negative interactions in irreversible self-assembly. Algorithmica **66**(1), 153–172 (2013)

22. Doty, D., Lutz, J.H., Patitz, M.J., Schweller, R.T., Summers, S.M., Woods, D.: The tile assembly model is intrinsically universal. In: Proceedings of the 53rd Annual Symposium on Foundations of Computer Science (FOCS), pp. 302–310 (2012)

23. Doty, D., Lutz, J.H., Patitz, M.J., Summers, S.M., Woods, D.: Intrinsic universality in self-assembly. In: Proceedings of International Symposium on Theoretical Aspects of Computer Science (STACS), LIPIcs, vol. 5, pp. 275–286. Schloss Dagstuhl (2010)

24. Doty, D., Patitz, M.J., Reishus, D., Schweller, R.T., Summers, S.M.: Strong fault-tolerance for self-assembly with fuzzy temperature. In: Foundations of Computer Science (FOCS), pp. 417–426 (2010)

25. Doty, D., Patitz, M.J., Summers, S.M.: Limitations of self-assembly at temperature one. In: Deaton, R., Suyama, A. (eds.) DNA 15. LNCS, vol. 5877, pp. 35–44. Springer, Heidelberg (2009)

26. Fekete, S.P., Hendricks, J., Patitz, M.J., Rogers, T.A., Schweller, R.T.: Universal computation with arbitrary polyomino tiles in non-cooperative self-assembly. In: Proceedings of the 26th Annual ACM-SIAM Symposium on Discrete Algorithms (SODA), pp. 148–167 (2015)

27. Freund, R., Kari, L., Pâun, G.: Dna computing based on splicing: the existence of universal computers. Theor. Comput. Syst. **32**, 69–112 (1999)

28. Goles, E., Meunier, P.-E., Rappaport, I., Theyssier, G.: Communication complexity and intrinsic universality in cellular automata. Theor. Comput. Sci. **412**(1–2), 2–21 (2011)

29. Hendricks, J., Patitz, M.J., Rogers, T.A.: The simulation powers and limitations of higher temperature hierarchical self-assembly systems. Technical report, arXiv (2015)

30. Jonoska, N., Karpenko, D.: Active tile self-assembly, part 1: universality at temperature 1. Int. J. Found. Comput. Sci. **25**(2), 141–163 (2014)

31. Luhrs, C.: Polyomino-safe DNA self-assembly via block replacement. Nat. Comput. **9**(1), 97–109 (2010)

32. Mañuch, J., Stacho, L., Stoll, C.: Two lower bounds for self-assemblies at temperature 1. J. Comput. Biol. **16**(6), 841–852 (2010)

33. Meunier, P.-E.: The self-assembly of paths and squares at temperature 1. Technical report, arXiv (2013)

34. Meunier, P.-E., Patitz, M.J., Summers, S.M., Theyssier, G., Winslow, A., Woods, D.: Intrinsic universality in tile self-assembly requires cooperation. In: Proceedings of the 25th Annual ACM-SIAM Symposium on Discrete Algorithms (SODA), pp. 752–771 (2014)

35. Ollinger, N.: Universalities in cellular automata. In: Rozenberg, G., Bäck, T., Kok, J.N. (eds.) Handbook of Natural Computing, pp. 190–229. Springer, Heidelberg (2012)

36. Padilla, J.E., Patitz, M.J., Pena, R., Schweller, R.T., Seeman, N.C., Sheline, R., Summers, S.M., Zhong, X.: Asynchronous signal passing for tile self-assembly: fuel efficient computation and efficient assembly of shapes. In: Mauri, G., Dennunzio, A., Manzoni, L., Porreca, A.E. (eds.) UCNC 2013. LNCS, vol. 7956, pp. 174–185. Springer, Heidelberg (2013)

37. Patitz, M.J.: An introduction to tile-based self-assembly. In: Durand-Lose, J., Jonoska, N. (eds.) UCNC 2012. LNCS, vol. 7445, pp. 34–62. Springer, Heidelberg (2012)

38. Patitz, M.J., Schweller, R.T., Summers, S.M.: Exact shapes and turing universality at temperature 1 with a single negative glue. In: Cardelli, L., Shih, W. (eds.) DNA 17 2011. LNCS, vol. 6937, pp. 175–189. Springer, Heidelberg (2011)

39. Pâun, G.: On the power of the splicing operation. Int. J. Comput. Math. **59**(1–2), 27–35 (1995)
40. Reif, J., Song, T.: The computation complexity of temperature-1 tilings. Technical report, Duke University (2014)
41. Rothemund, P.W.K., Winfree, E.: The program-size complexity of self-assembled squares. In: Proceedings of ACM Symposium on Theory of Computing (STOC), pp. 459–468 (2000)
42. Sahu, S., Reif, J.H.: Capabilities and limits of compact error resilience methods for algorithmic self-assembly in two and three dimensions. In: Mao, C., Yokomori, T. (eds.) DNA12. LNCS, vol. 4287, pp. 223–238. Springer, Heidelberg (2006)
43. Schulman, R., Winfree, E.: Programmable control of nucleation for algorithmic self-assembly. SIAM J. Comput. **39**(4), 1581–1616 (2009)
44. Soloveichik, D., Cook, M., Winfree, E.: Combining self-healing and proofreading in self-assembly. Nat. Comput. **7**(2), 203–218 (2008)
45. Soloveichik, D., Winfree, E.: Complexity of self-assembled shapes. In: Ferretti, C., Mauri, G., Zandron, C. (eds.) DNA 2004. LNCS, vol. 3384, pp. 344–354. Springer, Heidelberg (2005)
46. Summers, S.M.: Universality in algorithm self-assembly. Ph.D. thesis, Iowa State University (2010)
47. Winfree, E.: Algorithmic self-assembly of DNA. Ph.D. thesis, Caltech (1998)
48. Winfree, E.: Self-healing tile sets. In: Nanotechnology: Science and Computation, pp. 55–78. Springer, Heidelberg (2006)
49. Winfree, E., Bekbolatov, R.: Proofreading tile sets: Error correction for algorithmic self-assembly. In: Chen, J., Reif, J.H. (eds.) DNA 2003. LNCS, vol. 2943, pp. 126–144. Springer, Heidelberg (2004)
50. Woods, D.: Intrinsic universality and the computational power of self-assembly. Philos. Trans. Royal Soc. A **373**, 2015 (2046)
51. Yokomori, T., Kobayashi, S.: Dna-ec: a model of dna computing based on equality checking. In: DNA Based Computers III, DIMACS Series in Discrete Mathematics and Theoretical Computer Science, vol. 48, pp. 347–359. American Mathematical Society (1999)

Regular Papers

The Corona Limit of Penrose Tilings
Is a Regular Decagon

Shigeki Akiyama[1] and Katsunobu Imai[2]([⊠])

[1] Institute of Mathematics & Center for Integrated Research
in Fundamental Science and Technology, University of Tsukuba,
Tennodai, Tsukuba, Ibaraki 305-8571, Japan
akiyama@math.tsukuba.ac.jp
[2] Graduate School of Engineering, Hiroshima University,
Higashi-hiroshima 739-8527, Japan
imai@hiroshima-u.ac.jp

Abstract. We define and study the corona limit of a tiling, by investigating the signal propagations on cellular automata (CA) on tilings employing the simple growth CA. In particular, the corona limit of Penrose tilings is the regular decagon.

1 Introduction

Since the discovery of quasi-crystals, quasi-periodic tilings like Penrose tiling attracted a lot of attention as their possible mathematical models. Spectral study of tiling dynamical system and Schrödinger operator on quasi-periodic structure are developed in order to analyze their long-range order and quantum mechanical motion of the particle on quasi-periodic structure [8].

Cellular automata working on quasi-periodic tilings [2,11,16] are also studied. In particular, the intensive studies of the *Game of Life* [3] on Penrose tilings by Owens and Stepney are paid attentions [12,13]. As the result, recently, a cellular automata simulator for reaction-diffusion media which also working on a Penrose tiling is released [7] and the first glider pattern on Penrose tilings is also found [5].

We are interested in the difference of signal propagation of cellular automata working on between the normal periodic cells and quasi-periodic tilings. In this context, Chidyagwai and Reiter showed that the broken symmetry of quasi-periodic tilings, while still retaining a highly organized structure, could be used to simulate the complex growth of snow crystals [2]. They could produce global n-fold symmetry models where regular hexagonal grids could only produce 6-fold symmetry models.

In this paper, we introduce **corona limit** which naturally visualizes the growth pattern of signal propagation. We show that the corona limit of a Penrose tilings is a regular decagon. The speed of convergence depends on the version of tilings and adjacency condition of it.

© IFIP International Federation for Information Processing 2016
Published by Springer International Publishing Switzerland 2016. All Rights Reserved
M. Cook and T. Neary (Eds.): AUTOMATA 2016, LNCS 9664, pp. 35–48, 2016.
DOI: 10.1007/978-3-319-39300-1_4

2   Corona Limit

A **tiling** $\mathcal{T}$ is a covering of $\mathbb{R}^2$ by finitely many polygonal tiles and their images by isometry (translation, rotation, and flip) which overlap only at their boundaries. Two tiles A, B are adjacent (resp. edge adjacent) if they share a point (resp. an edge). For brevity, we also say that A and A itself are adjacent (edge adjacent) as well. A **patch** $\mathcal{P}$ is a finite set of tiles in $\mathcal{T}$. The 1-st **corona** $\mathcal{P}^{(1)}$ of $\mathcal{P}$ is a patch consisting of all tiles which is adjacent to a tile of $\mathcal{P}$. The n-th corona $\mathcal{P}^{(n)}$ is defined as a corona of the $(n-1)$-th corona $\mathcal{P}^{(n-1)}$ for $n = 2, 3, \dots$. An edge corona $\mathcal{P}^{[1]}$ and n-th edge corona $\mathcal{P}^{[n]}$ are defined in the same manner but by edge adjacency. The concept of corona is important and well-known, which is used in the basic theory of tilings, e.g. [6, Chap 3.2], [15] and [14]. It also appears in Heesch's problem ([6, Chap 3.8.3] and [10]) which forms a counter part of existence problem of tilings by a given set of tiles. If a sequence of shrunk patches

$$\frac{1}{n}\mathcal{P}^{(n)} \qquad n = 1, 2, \dots$$

converges to a non-empty compact set K under Hausdorff metric, we say that K is a **corona limit**. Here for two non empty compact sets A and B, the Hausdorff metric is defined as

$$\inf\{\varepsilon > 0 \mid A \subset B[\varepsilon] \text{ and } B \subset A[\varepsilon]\}$$

with $X[\varepsilon] := \{y \in \mathbb{R}^2 \mid \exists x \in X \; \|y - x\| \leq \varepsilon\}$. To see the geometric meaning better, the limit is rephrased as $\lim_n \frac{1}{n}\left(\mathcal{P}^{(n)} - c\right)$ by fixing a point c in the initial patch $\mathcal{P}$. Then the coronas grow around the center c, and we renormalize them by the factor $1/n$ to obtain the corona limit. Since the tiling has finitely many shapes, the diameter and inradius (maximal radius of the inscribed ball) of the n-th corona are bounded from below and above by positive constant multiples of n. Thus the corona limit contains the origin as an inner point, and is bounded. Interestingly, one can show that if a corona limit K exists, then it does not depend on the choice of the initial patch $\mathcal{P}$. To see this, we claim that the corona limit of $\mathcal{P}$ and that of $\mathcal{P}^{(n)}$ are identical, and for any two patches $\mathcal{P}, \mathcal{Q}$, there exist a positive integer m that $\mathcal{Q} \subset \mathcal{P}^{(m)}$ and $\mathcal{P} \subset \mathcal{Q}^{(m)}$. Therefore we say that K is the corona limit of $\mathcal{T}$. We can similarly define the edge corona limit. For periodic tilings, the corona limit is usually easy to obtain by simple induction. For the standard square tiling the corona limit is a square, and the edge corona limit is also a square but rotated $\pi/4$. It often becomes a hexagon for Archimedean tilings.

3   Penrose Tilings

A Penrose tiling is a tiling generated by a set of tiles with matching conditions. They tiles the plane but only in non periodic way. Among many versions of Penrose tilings, there are two types of tilings generated by two quadrilateral proto-tiles : two rhombus tiles (a *fat* and a *thin*), and a *kite* and a *dart*. To avoid periodic arrangements, several matching conditions are known to be added on these tiles [4]. Ammann bars are one of such matching condtions [6]. Each tile in Fig. 1 has Ammann bar line segments and Fig. 2 illustrates a possible rhombus tiling and its substitution rules. Let φ be the golden ratio. To get a tiling from a patch, iterate the magnification by φ and substitution for each tile by the rule in Fig. 2. The line segments of each tile must be continued straight across the boundary. They form parallel lines of five different slopes and the gap length of each parallel lines is one of L and S. It is known that there exist infinitely many translationaly inequivalent quasi-periodic tilings of the plane with these tiles.

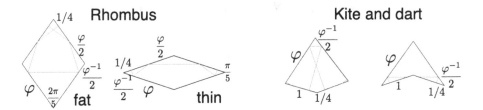

Fig. 1. *Rhombus*, and *kite and dart* tiles with the line segments of Ammnann bars.

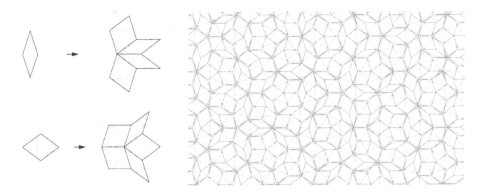

Fig. 2. A rhombus tiling and its substitution rules.

4   Growth Cellular Automaton on Penrose Tilings

We define a cellular automaton $\mathcal{A}$ as follows: choose a two-dimensional tiling $\mathcal{T}$. Each tile τ has a state $c(\tau) \in Q$ and neighborhood tiles are defined by $\mathcal{N}(\tau)$. We call all assignments of state to the tiles $\mathcal{C} : \mathcal{T} \rightarrow Q$ as the configuration of $\mathcal{A}$. The local function of $\mathcal{A}$ is defined by the assignment of states in $\mathcal{N}(\tau)$ and returns the next state of τ. Thus its simultaneous application to each tile in $\mathcal{C}$ defines the global evolution $\mathcal{A} : \mathcal{C} \rightarrow \mathcal{C}$. When the tiling is the regular square grid, we denote the vertex adjacent neighborhood by Moore neighborhood ($\mathcal{N}_M$) and the edge adjacent neighborhood by von Neumann neighborhood ($\mathcal{N}_N$).

In order to define a CA on a Penrose tiling, we need to extend the above neighboring relation. Because it is not a lattice tiling, congruent tiles may have different neighborhoods. Generalized von Neumann and Moore neighborhoods are illustrated in Figs. 3 and 4, respectively [12].

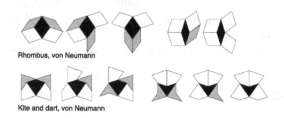

Fig. 3. The generalized von Neumann neighborhoods on Penrose tilings [12]

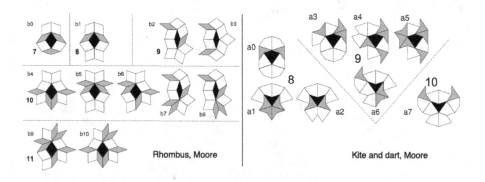

Fig. 4. The generalized Moore neighborhoods on Penrose tilings [12]

From this section, we focus on a simple cellular automaton, *growth cellular automaton.*

Definition 1. *Growth cellular automaton is a cellular automaton $\mathcal{A}$ whose state set is $Q = \{0, 1\}$. Its local function is defined such that if the state of a focus tile is 0 and at least one of its neighborhood tiles' state is 1 then its state changes to 1. Any tile of state 1 never changes its state.*

Let C_0 be a configuration such that a single cell's state is 1 and the others are 0. Let P_n be the shape formed by state 1 cells by n-step evolutions of $\mathcal{A}$ from C_0 as its initial configuration.

It is clear that in the case of square Moore neighborhood, P_n is a square which side length is $2n + 1$ and in the case of square von Neumann neighborhood, P_n is a $\pi/4$-rotated square which diagonal length is $2n + 1$.

We show the simulated results of the case of Penrose tilings. Figure 5 illustrates the results after 10-step and 30-step evolutions (The scale of the figures of

10 and 20 steps are different). Because the $|\mathcal{N}_M|$ is larger than $|\mathcal{N}_N|$ on average, the size of P_n is larger in the case of Moore neighborhoods than the case of von Neumann neighborhood. Each P_n seems to converge to a regular decagonal shape. Even the case of rhombus Moore neighborhood, it eventually converges to a regular decagonal shape (See Sect. 6).

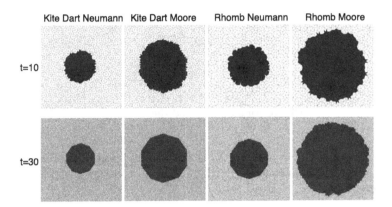

Fig. 5. The evolution of growth CA $\mathcal{A}$ on Penrose tilings ($t = 10, 30$ steps).

In the next section we show its proof in the case of rhombus von Neumann neighborhood.

5 Rhombus von Neumann Neighborhood Case

In this section, we prove the experimental observation of the previous section in the case of rhombus von Neumann neighborhood.

Choose a patch of the star shape in any rhombus Penrose tiling $\mathcal{T}$ as in Fig. 6. It has five symmetric crossing Ammann bars. We denote the central point of the star by O. We call the five symmetric Ammann bars across the star as a_0, b_0, c_0, d_0, e_0. We denote by ε the distance between O and one of the Ammann bars. We denote by X_i for each Ammann bar parallel to X_0 where $X \in \{a, b, c, d, e\}$ and for all integers i. We also denote X_{-i} by $\bar{X}_i$. The length of the gap between two Ammann bars X_i and X_{i+1} (denoted by $\overline{X_i X_{i+1}}$) is L or S. We denote by $X_i Y_j$ the cross point of two non-parallel Ammann bars X_i and Y_j.

Proposition 1. *Let ∂D_i be connected lines formed by the points: $a_i b_i$, $b_i c_i$, $c_i d_i$, $d_i e_i$, $e_i \bar{a}_i$, $\bar{a}_i \bar{b}_i$, $\bar{b}_i \bar{c}_i$, $\bar{c} \bar{d}_i$, $\bar{d} \bar{e}_i$, $\bar{e}_i a_i$, $a_i b_i$. There exists a constant k, for any $i (> k)$, ∂D_i forms a decagon (D_i) and $D_i \subset D_{i+1}$.*

Proof. We denote by $\hat{X}_i$ the distance between O and X_i. For any two Ammann bars X_i and Y_i which angle is $\pi/5$, if $\hat{X}_i / \hat{Y}_i > \cos(\pi/5)$ then ∂D_i forms a decagon.

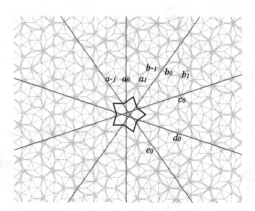

Fig. 6. Amman bars and their indices.

Because $\hat{X}_i = \overline{X_0, X_i} \pm \varepsilon$ and $\overline{X_0, X_i}$ is the i-th addition of L or S, the value $\hat{X}_i/\hat{Y}_i$ converges to 1 as i grows. So there exists a constant k, for any $i(> k)$, $\hat{X}_i/\hat{Y}_i > \cos(\pi/5)$.

We call D_i (where i satisfies that each D_i forms a decagon) as a sequence of *uniaxial Ammann bar decagons* (Fig. 7).

Proposition 2. *Decagon D_i converges to the shape of regular decagon as i goes to infinity.*

Proof. Because the ratio of the numbers of L and S for parallel Ammann bars converges to φ (10.6.8 in [6,9]), i.e., for any large i, the distance between O and X_i converges to the same length. Thus D_i converges to a regular decagon as i goes to infinity.

Definition 2. *If the part of the area of a tile τ in the gap between two Ammann bars X_i and X_{i+j} is larger than the remaining part, then we call τ is a gap tile of X_i and X_{i+j}. We denote by $G(X_i, X_{i+j})$ all gap tiles of X_i and X_{i+j}.*

Definition 3. *$G(X_i, X_{i+1})$ is said to be filled when the state of all tiles in $G(X_i, X_{i+1})$ is 1 except fat tiles of which longer diagonal lines are perpendicular to the bar X_i or X_{i+1}. We do not concern the states of these fat tiles.*

Figure 8 shows a filled S-gap and a filled L-gap.

Proposition 3. *For any tile $\tau \in G(X_i, X_{i+1})$, the following properties hold:*

1. *τ intersects only one of two Ammann bars X_i and X_{i+1};*
2. *If τ intersects X_i (X_{i+1}) then there exists an edge adjacent tile in $G(X_i, X_{i+1})$ which intersects the bar X_{i+1} (X_i) and τ has no adjacent tile in $G(X_{i+1}, X_{i+2})$ ($G(X_{i-1}, X_i)$);*
3. *If τ intersects X_i (X_{i+1}), then τ has an edge adjacent tile in $G(X_{i-1}, X_i)$ ($G(X_{i+1}, X_{i+2})$).*

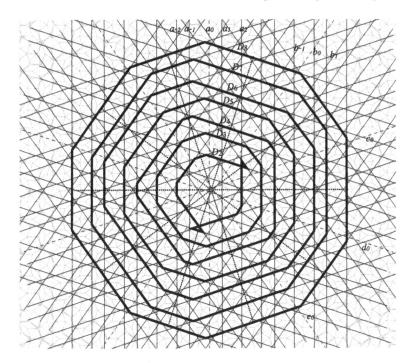

Fig. 7. A sequence of uniaxial Ammann bar decagons $D_i (i \geq 3)$

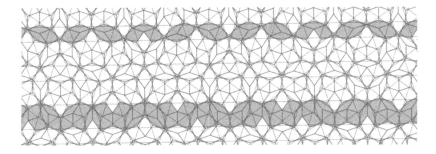

Fig. 8. S- and L-filled gaps.

Proof. There are 16 possible combinations of two tiles. Removing rotationally symmetric pairs, there are only eight pairs in Fig. 9. Moreover one of them is an *illegal* combination, i.e., if the pattern is included, some of its Ammann bar segments cannot form lines. The pairs L1 (S1, S2) is the component of two pairs of parallel Ammann bars with L-gaps (S-gaps), respectively. The pairs L2 and L3 are the component of one pair of parallel Ammann bars with L-gaps and they are symmetric with each other. The pairs LS1 and LS2 are the component of parallel Ammann bars with an L-gap and an S-gap and they are also symmetric with each other. Dashed (dotted) line segments are used for L-gaps (S-gaps), respectively.

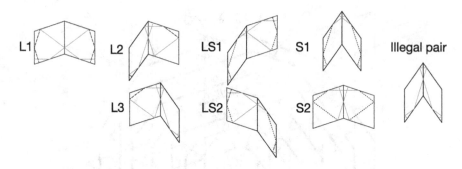

Fig. 9. Possible pairs according to Ammann bars.

To exhaust the possible placements along an S-gap, we start from a pair S1. Although S1 contains line segments for two distinct S-gaps, they are symmetric. So it is enough to consider one pair of them. Figure 10(1) shows the first possible extensions. The right patch in Fig. 10(1) is the only possibility because the left patch contains some Ammann bar segments which cannot form limes.

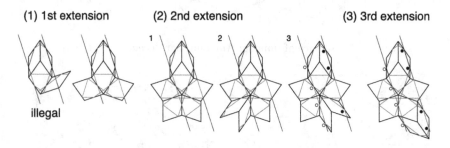

Fig. 10. Possible extensions from a pair S1.

The second step, there are three possibilities in Fig. 10(2). S1 appears again in Fig. 10(2)-3. Figure 10(2)-1,(2)-2 have a fat tile of the same angle along the S-gap. But the latter is an illegal placement because there is no feasible placement in the next extension. Thus we can continue the next step only in the case of 1. The final step, the only possible placement is Fig. 10(3) and S1 appears again.

Figure 11 is the possible extensions to the opposite direction from a pair S1. Because any thin tile cannot be placed to the opposite side of S1 as in Fig. 11(1), fat tiles forced to be placed next to S1. Figure 11(2)-1, 2 have already appeared in Fig. 10(2)-1, 2, respectively. S1 appears again in Fig. 11(2)-3. Thus there is no different placement except their symmetric cases appeared in the both side of S1.

The downside (upperside) of S2 appears in Fig. 10(1) right (Fig. 10(2)), respectively. The upper side of LS1 (and the symmetric version, LS2) appears in Fig. 10(1) right. The downside of LS1 (and the symmetric version, LS2) only

(1) 1st extension   **(2) 2nd extension**

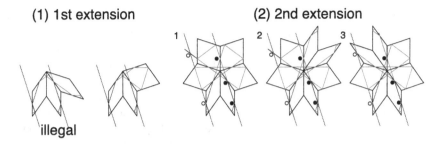

Fig. 11. Possible extensions of a tile to the opposite direction from a pair S1.

accepts a thin tile to form the shape S1. Thus there is no more S-gap patterns connected from S1, S2, LS1, and LS2.

Because most of the patterns forming an S-gap are S1, S2, LS1, and LS2, the other patterns appeared in an S-gap must be generated by the above process. So all possible placements except their symmetric cases along an S-gap are included.

In the same way, we consider the possible placements along an L-gap, we start from a pair L1. Although L1 contains line segments for two distinct L-gaps, they are symmetric. So it is enough to consider one pair of them. Figure 12 (1) shows the first possible extensions. There is only one feasible placement and the next extensions of it are in Fig. 12 (2).

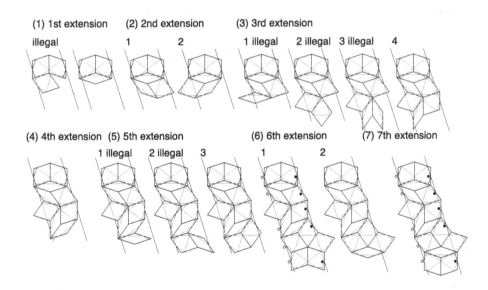

Fig. 12. The possible extensions from a pair L1.

We consider the next step of Fig. 12 (2)-1. The only feasible placement is Fig. 12 (3)-4 and a thin tile in the next step (Fig. 12 (4)). Thus we do not need to consider the case of Fig. 12 (2)-2 separately, because the angle of the thin tile in Fig. 12 (4) is the same. The next step, only the case, Fig. 12 (5)-3, is possible and there are two options Fig. 12 (6) after the extension. L1 appears again in Fig. 12 (6)-1. The final extension in Fig. 12 (7) shows the only legal extension and the opposite side of L1 appears.

Then we consider the opposite side of L1. Figure 13 is the possible extensions to the opposite direction from a pair L1. In Fig. 13-1, it appears the opposite side of L1 again and Fig. 13-2 (Fig. 13-3) has the same shape as Fig. 12 (2)-1 (Fig. 12 (2)-2), respectively. Thus all possible patterns appeared from L1 are shown in the previous figures.

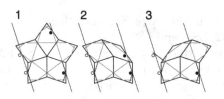

Fig. 13. Possible extension to the opposite direction from a pair L1.

The upper side of L2 (and its symmetric version) appears in Fig. 12 (2). The downside of L2 (and its symmetric version) appears in Fig. 12 (5)-3. The downside of LS1 (and the symmetric version, LS2) appears in Fig. 12 (2)-2. The upper side of LS1 (and the symmetric version, LS2) only accepts a thin tile and succeeding placements are the same as that in Fig. 12 (4). Thus there is no more L-gap patterns connected from L1, L2, L3, LS1, and LS2.

To sum up, all possible placement of tiles along S-gaps and L-gaps are included in the above figures. Thus it is clear that the proposition agree with the all patterns in the figures.

Proposition 3 implies that a state-1 signal coming to a side of a gap $G(X_i, X_{i+1})$ takes exactly two steps to pass through the gap. Thus the following proposition holds:

Proposition 4. *Let $\mathcal{A}$ be a von Neumann neighborhood growth cellular automaton on a rhombus Penrose tiling. All tiles of its initial configuration are state 0 except a filled gap of two Ammann bars X_i and X_{i+1}. After two steps execution of $\mathcal{A}$, $G(X_{i+1}, X_{i+2})$ and $G(X_{i-1}, X_i)$ are filled and all gaps outside of X_{i+2} or X_{i-1} are remained to be unfilled.*

By Proposition 4, a sufficiently long linear wavefront of state-1 cells parallel to an Ammann bar preserve its linearity after filling the next gap. But we need to show the case of decagon. In the following, we show how each node of a filled decagon D_{k+1} is formed by its previous filled decagon D_k.

Definition 4. *Let D_i be a sequence of uniaxial Ammann bar decagons. For an integer k, we call D_k is filled when the state of all tiles in D_k is 1, except fat tiles of which longer diagonal lines are perpendicular and crossing with each line in ∂D_k. We do not concern the states of these fat tiles.*

Proposition 5. *Let $\mathcal{A}$ be a von Neumann neighborhood growth cellular automaton on a rhombus Penrose tiling and D_i be a sequence of uniaxial Ammann bar decagons. Suppose all tile has state 0 in its initial configuration except a filled decagon D_k for an integer k. Then after two-step executions of $\mathcal{A}$, D_{k+1} is filled and the outside of D_{k+1} is remained to be unfilled.*

Proof. By Proposition 4, the outside of D_{k+1} is remained to be unfilled. We need to show that all tiles in $G(X_k, X_{k+1}) \cap D_{k+1}$ are filled after two-step executions. It is enough to show that the closest tile (in D_{k+1}) to each vertex of D_{k+1} can be reached in two-step executions from its associated vertex of D_k. Two crossing points of two pairs of two neighboring parallel Ammann bars of which angle is $4\pi/5$ are the candidates of vertices of D_k and D_{k+1}. In the pictures of proof of Proposition 3, black circles and white circles are all such candidates of points. It is easy to check that they all have two-step distance.

Theorem 1. *The edge corona limit of rhombus Penrose tilings is a regular decagon.*

Proof. Let $\mathcal{A}$ be a von Neumann neighborhood growth cellular automaton on a rhombus Penrose tiling with any finite initial configuration (the number of state-1 cell is a finite natural number). We can choose a sequence of uniaxial Ammann bar decagons D_i close to the initial configuration. Even if the given initial configuration is not connected, executing enough steps of $\mathcal{A}$, it is possible to change the shape of state-1 cells as follows:

1. the shape formed by the state 1 tiles is connected;
2. the shape contains a star and at least one of Ammann bar decagon D_i for some i,

i.e., there exists integers i and $j(> i)$ such that D_i is filled and tiles outside of D_j are state 0. We denote this initial patch by $\mathcal{P}^{[0]}$. By Proposition 5, D_{n+i} is filled and tiles outside of D_{n+j} are state 0 after $2n$-step executions, i.e., $D_{n+i} \subset \mathcal{P}^{[2n]} \subset D_{n+j}$ holds. Because $j - i$ is a constant, $\mathcal{P}^{[n]}/n$ converges to the regular decagon $\lim_{n \to \infty} D_{[n/2]}/n$ as n tends to infinity.

6 The Difference of Growth Speeds

In the previous section, we show the wavefront of a growth cellular automaton on a rhombus Penrose tilings eventually forms a regular decagon and the speed of the wavefront passing through L- or S-gap is $l/2$ per step, $l \in \{S, L\}$. Although the signal propagation of the Moore neighborhood case is more fluctuating, the wavefront also forms a regular decagon and its speed is l, $l \in \{S, L\}$ by the existence of thin tiles along an S-gap. In the case of kite and dart tilings, it is also possible to show the similar result in the same way.

Each growth speed across a group of Ammnann bars are shown in Table 1. They actually agree with the difference of growth illustrated in Fig. 5.

Table 1. Growth speeds.

	rhombus	kite and dart
von Neumann	$L/2$, $S/2$	$L/3$, $S/2$
Moore	L, S	$L/2$, S

In Sect. 4, we pointed out that the convergence speed to a decagonal shape is very slow in the case of rhombus Moore neighborhood. To inspect the behavior, we modify the local function of the growth cellular automaton as follows:

Definition 5. *A cellular automaton $\mathcal{A}_+$ whose state set is $Q = \{0, 1, 2, 3, \dots\}$ ($|Q|$ is the maximum number of tiles sharing a vertex). Its local function is defined such that if the state of a focus tile is 0 then its state is changed to the sum of all non-zero cells in $\mathcal{N}$. Any tile with state ≥ 1 never changes its state.*

Figure 14 illustrates the evolutions of $\mathcal{A}_+$ in the case of square and rhombus Penrose tilings. A light gray (state 1) tile receives a signal from a neighboring

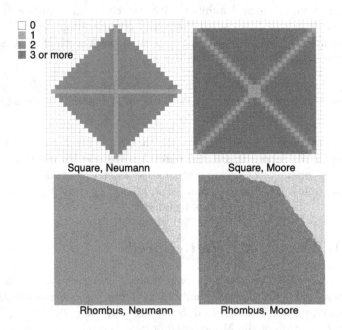

Fig. 14. The evolutions of $\mathcal{A}_+$

tile and a darker tile receives two or more signals at a time from two or more neighboring tiles. Intuitively a light gray line segment shows that the existence of the fastest signal propagation at the sites. In the square cases, the line segments are the diagonal lines of a square formed by the 'light speed' of the cellular space.

In the rhombus cases, there are many light gray line segments of short length. This mainly due to the fluctuation of signal propagations caused by the existence of alternations of L and S gaps. But even the tiling is the same, the situation is quite different between the von Neumann and the Moore neighborhood cases. In the Moore neighborhood case, there are many self similar light gray line segments and complicated signal collisions occur. This is the reason for the slow convergence to a regular decagonal shape in the viewpoint of signal propagation of cellular automata.

As the result, employing rhombus tiling and Moore neighborhood might be more suitable for the cellular automata simulation of, for example, chemical reactions than the other cases. Because it anyway achieves the most homogeneous local signal propagation in the above options.

7   Conclusion

In this paper, we showed that the corona limit of Penrose tilings is a regular decagon, thanks to the existence of Ammann bars. This might suggest some relation between corona limits and windows of the cut and project scheme which generate the tilings. However we know little on this connection. This idea may fail if the cut and project scheme is non Archimedean. For the case of the chair tiling (Fig. 10.1.5 in [6]), its corona limit is easily shown to be an octagon. The associated cut and project scheme is realized with 2-adic internal space [1] whose window is not a four dimensional hypercube.

It is an intriguing question to understand what decides the shape of the corona limit.

Acknowledgments. Shigeki Akiyama is supported by the Japanese Society for the Promotion of Science (JSPS), Grant in aid 21540012. Katsunobu Imai thanks Takahiro Hatsuda from SOLPAC Co., Ltd. for the helpful discussions and gratefully acknowledges the support of JSPS KAKENHI Grant Number 26330016.

References

1. Baake, M., Moody, R.V., Schlottmann, M.: Limit-(quasi)periodic point sets as quasicrystals with p-adic internal spaces. J. Phys. A: Math. Gen. **31**, 27 (1988)
2. Chidyagwai, P., Reiter, C.A.: A local cellular model for growth on quasicrystals. Chaos, Solutions Fractals **24**, 803–812 (2005)
3. Gardner, M.: Mathematical games - the fantastic combinations of john conway's new solitaire game "life". Sci. Am. **223**, 120–123 (1970)
4. Gardner, M.: Extraordinary nonperiodic tiling that enriches the theory of tiles. Sci. Am. **236**, 110–121 (1977)

5. Goucher, A.P.: Gliders in cellular automata on penrose tilings. J. Cell. Automata **7**(5–6), 385–392 (2012)
6. Grünbaum, B., Shephard, C.G.: Tiling and Patterns. Freeman, San Francisco (1987)
7. Hutton, T.: Ready: a cross-platform implementation of various reaction-diffusion systems, https://github.com/GollyGang/ready
8. Kellendonk, J., Lenz, D. and Savinien, J. (ed.), Mathematics of aperiodic order, Progress in Mathematical Physics, vol. 309. p. xi+428. Birkhuser/Springer, Basel (2015)
9. Lück, R.: Basic ideas of ammann bar grids. Int. J. Mod. Phys. B **7**(6 & 7), 1437–1453 (1993)
10. Mann, C.: Heesch's tiling problem. Am. Math. Mon. **111**(6), 509–517 (2004)
11. McClure, M.: A stohastic cellular automaton for three-coloring penrose tiles. Comput. Graph. **26**(3), 519–524 (2002)
12. Owens, N., Stepney, S.: Investigation of the game of life cellular automata rules on penrose tilings: lifetime, ash and oscillator statistics. J. Cell. automata **5**(3), 207–225 (2010)
13. Owens, N., Stepney, S.: The game of life rules on penrose tilings: still life and oscillators. In: Adamatzky, A. (ed.) Game of Life Cellular Automata, pp. 331–378. Springer, London (2010)
14. Schattschneider, D., Dolbilin, N.: One corona is enough for the euclidean plane. In: Patera, J. (ed.) Quasicrystals and Discrete Geometry. Fields Institute Monographs, vol. 10, pp. 193–199. AMS, Providence (1998)
15. Solomyak, B.: Nonperiodicity implies unique composition for self-similar translationally finite tilings. Discrete Comput. Geom. **20**(2), 265–279 (1998)
16. Wolfram, S.: A New Kind of Science. Wolfram Media, Champaign (2002). p. 929

The Group of Reversible Turing Machines

Sebastián Barbieri[1(✉)], Jarkko Kari[2], and Ville Salo[3]

[1] LIP, ENS de Lyon – CNRS – INRIA – UCBL – Université de Lyon, Lyon, France
sebastian.barbieri@ens-lyon.fr
[2] University of Turku, Turku, Finland
jkari@utu.fi
[3] Center for Mathematical Modeling, University of Chile, Santiago, Chile
vosalo@utu.fi

Abstract. We consider Turing machines as actions over configurations in $\Sigma^{\mathbb{Z}^d}$ which only change them locally around a marked position that can move and carry a particular state. In this setting we study the monoid of Turing machines and the group of reversible Turing machines. We also study two natural subgroups, namely the group of finite-state automata, which generalizes the topological full groups studied in the theory of orbit-equivalence, and the group of oblivious Turing machines whose movement is independent of tape contents, which generalizes lamplighter groups and has connections to the study of universal reversible logical gates. Our main results are that the group of Turing machines in one dimension is neither amenable nor residually finite, but is locally embeddable in finite groups, and that the torsion problem is decidable for finite-state automata in dimension one, but not in dimension two.

1 Introduction

1.1 Turing Machines and Their Generalization

Turing machines have been studied since the 30 s as the standard formalization of the abstract concept of computation. However, more recently, Turing machines have also been studied in the context of dynamical systems. In [20], two dynamical systems were associated to a Turing machine, one with a 'moving tape' and one with a 'moving head'. After that, there has been a lot of study of dynamics of Turing machines, see for example [1,11,12,15,16,19,28]. Another connection between Turing machines and dynamics is that they can be used to define subshifts. Subshifts whose forbidden patterns are given by a Turing machine are called effectively closed, computable, or Π^0_1 subshifts, and especially in multiple dimensions, they are central to the topic due to the strong links known between SFTs, sofic shifts and Π^0_1-subshifts, see for example [4,9]. An intrinsic notion of Turing machine computation for these subshifts on general groups was proposed in [3], and a similar study was performed with finite state machines in [26,27].

In all these papers, the definition of a Turing machine is (up to notational differences and switching between the moving tape and moving head model) the

© IFIP International Federation for Information Processing 2016
Published by Springer International Publishing Switzerland 2016. All Rights Reserved
M. Cook and T. Neary (Eds.): AUTOMATA 2016, LNCS 9664, pp. 49–62, 2016.
DOI: 10.1007/978-3-319-39300-1_5

following: A Turing machine is a function $T : \Sigma^{\mathbb{Z}} \times Q \to \Sigma^{\mathbb{Z}} \times Q$ defined by a local rule $f_T : \Sigma \times Q \to \Sigma \times Q \times \{-1, 0, 1\}$ by the formula

$$T(x, q) = (\sigma_{-d}(\tilde{x}), q') \text{ if } f_T(x_0, q) = (a, q', d),$$

where $\sigma : \Sigma^{\mathbb{Z}} \to \Sigma^{\mathbb{Z}}$ is the shift action given by $\sigma_d(x)_z = x_{z-d}$, $\tilde{x}_0 = a$ and $\tilde{x}|_{\mathbb{Z} \setminus \{0\}} = x|_{\mathbb{Z} \setminus \{0\}}$. In this paper, such Turing machines are called *classical Turing machines*. This definition (as far as we know) certainly suffices to capture all computational and dynamical properties of interest, but it also has some undesirable properties: The composition of two classical Turing machines – and even the square of a classical Turing machine – is typically not a classical Turing machine, and the reverse of a reversible classical Turing machine is not always a classical Turing machine.

In this paper, we give a more general definition of a Turing machine, by allowing it to move the head and modify cells at an arbitrary (but bounded) distance on each timestep. With the new definition, we get rid of both issues: With our definition,

– Turing machines are closed under composition, forming a monoid, and
– reversible Turing machines are closed under inversion, forming a group.

We also characterize reversibility of classical Turing machines in combinatorial terms, and show what their inverses look like. Our definition of a Turing machine originated in the yet unpublished work [25], where the group of such machines was studied on general $\mathbb{Z}$-subshifts (with somewhat different objectives).

These benefits of the definition should be compared to the benefits of allowing arbitrary radii in the definition of a cellular automaton: If we define cellular automata as having a fixed radius of, say, 3, then the inverse map of a reversible cellular automaton is not always a cellular automaton, as the inverse of a cellular automaton may have a much larger radius [8]. Similarly, with a fixed radius, the composition of two cellular automata is not necessarily a cellular automaton.

We give our Turing machine definitions in two ways, with a moving tape and with a moving head, as done in [20]. The moving tape point of view is often the more useful one when studying one-step behavior and invariant measures, whereas we find the moving head point of view easier for constructing examples, and when we need to track the movement of multiple heads. The moving head Turing machines are in fact a subset of cellular automata on a particular kind of subshift. The moving tape machine on the other hand is a generalization of the topological full group of a subshift, which is an important concept in particular in the theory of orbit equivalence. For topological full groups of minimal subshifts, see for example [13,14,17]. The (one-sided) SFT case is studied in [24].

1.2   Our Results and Comparisons with Other Groups

In Sect. 2, we define our basic models and prove basic results about them. In Sect. 2.3, we define the uniform measure and show as a simple application of it that injectivity and surjectivity are both equal to reversibility.

Our results have interesting counterparts in the theory of cellular automata: One of the main theorems in the theory of cellular automata is that injectivity implies surjectivity, and (global) bijectivity is equivalent to having a cellular automaton inverse map. Furthermore, one can attach to a reversible one- or two-dimensional cellular automaton its 'average drift', that is, the speed at which information moves when the map is applied, and this is a homomorphism from the group of cellular automata to a sublattice of $\mathbb{Q}^d$ (where d is the corresponding dimension), see [18]. In Sect. 3 we use the uniform measure to define an analog, the 'average movement' homomorphism for Turing machines.

In Sect. 3, we define some interesting subgroups of the group of Turing machines. First, we define the local permutations – Turing machines that never move the head at all –, and their generalization to oblivious Turing machines where movement is allowed, but is independent of the tape contents. The group of oblivious Turing machines can be seen as a kind of generalization of lamplighter groups. It is easy to show that these groups are amenable but not residually finite. What makes them interesting is that the group of oblivious Turing machines is finitely generated, due to the existence of universal reversible logical gates. It turns out that strong enough universality for reversible gates was proved only recently [2].

We also define the group of (reversible) finite-state machines – Turing machines that never modify the tape. Here, we show how to embed a free group with a similar technique as used in [10], proving that this group is non-amenable. By considering the action of Turing machines on periodic points,[1] we show that the group of finite-state automata is residually finite, and the group of Turing machines is locally embeddable in finite groups (in particular sofic).

Our definition of a Turing machine can be seen as a generalization of the topological full group, and in particular finite-state machines with a single state exactly correspond to this group. Thus, it is interesting to compare the results of Sect. 3 to known results about topological full groups. In [14,17] it is shown that the topological full group of a minimal subshift is locally embeddable in finite groups and amenable, while we show that on full shifts, this group is non-amenable, but the whole group of Turing machines is LEF.[2]

Our original motivation for defining these subgroups – finite-state machines and local permutations – was to study the question of whether they generate all reversible Turing machines. Namely, a reversible Turing machine changes the tape contents at the position of the head and then moves, in a globally reversible way. Thus, it is a natural question whether every reversible Turing machine can actually be split into reversible tape changes (actions by local permutations) and reversible moves (finite-state automata). We show that this is not the case, by showing that Turing machines can have arbitrarily small average movement,

---

[1] The idea is similar as that in [23] for showing that automorphism groups of mixing SFTs are residually finite, but we do not actually look at subsystems, but the periodic points of an enlarged system, where we allow infinitely many heads to occur.

[2] In [25] it is shown that on minimal subshifts, the group of Turing machines coincides with the group of finite-state automata.

but that elementary ones have only a discrete sublattice of possible average movements. We do not know whether this is the only restriction.

In Sect. 4, we show that the group of Turing machines is recursively presented and has a decidable word problem, but that its torsion problem (the problem of deciding if a given element has finite order) is undecidable in all dimensions. For finite-state machines, we show that the torsion problem is decidable in dimension one, but is undecidable in higher dimensions, even when we restrict to a finitely generated subgroup. We note a similar situation with Thompson's group V: its torsion problem is decidable in one-dimension, but undecidable in higher dimensions [5, 6].

1.3   Preliminaries

In this section we present general definitions and settle the notation which is used throughout the article. The review of these concepts will be brief and focused on the dynamical aspects. For a more complete introduction the reader may refer to [22] or [7] for the group theoretic aspects. Let $\mathcal{A}$ be a finite alphabet. The set $\mathcal{A}^{\mathbb{Z}^d} = \{x : \mathbb{Z}^d \to \mathcal{A}\}$ equipped with the left group action $\sigma : \mathbb{Z}^d \times \mathcal{A}^{\mathbb{Z}^d} \to \mathcal{A}^{\mathbb{Z}^d}$ defined by $(\sigma_v(x))_u = x_{u-v}$ is a *full shift*. The elements $a \in \mathcal{A}$ and $x \in \mathcal{A}^{\mathbb{Z}^d}$ are called *symbols* and *configurations* respectively. With the discrete topology on $\mathcal{A}$ the set of configurations $\mathcal{A}^{\mathbb{Z}^d}$ is compact and given by the metric $d(x, y) = 2^{-\inf(\{|v| \in \mathbb{N} |\ x_v \neq y_v\})}$ where $|v|$ is a norm on $\mathbb{Z}^d$ (we settle here for the $|| \cdot ||_\infty$ norm). This topology has the *cylinders* $[a]_v = \{x \in \mathcal{A}^{\mathbb{Z}^d} | x_v = a \in \mathcal{A}\}$ as a subbasis. A *support* is a finite subset $F \subset \mathbb{Z}^d$. Given a support F, a *pattern with support F* is an element p of $\mathcal{A}^F$. We also denote the cylinder generated by p in position v as $[p]_v = \bigcap_{u \in F} [p_u]_{v+u}$, and $[p] = [p]_0$.

Definition 1. *A subset X of $\mathcal{A}^{\mathbb{Z}^d}$ is a subshift if it is σ-invariant $- \sigma(X) \subset X -$ and closed for the cylinder topology. Equivalently, X is a subshift if and only if there exists a set of forbidden patterns $\mathcal{F}$ that defines it.*

$$X = \bigcap_{p \in \mathcal{F}, v \in \mathbb{Z}^d} \mathcal{A}^{\mathbb{Z}^d} \setminus [p]_v.$$

Let X, Y be subshifts over alphabets $\mathcal{A}$ and $\mathcal{B}$ respectively. A continuous $\mathbb{Z}^d$-equivariant map $\phi : X \to Y$ between subshifts is called a morphism. A well-known Theorem of Curtis, Lyndon and Hedlund which can be found in full generality in [7] asserts that morphisms are equivalent to maps defined by local rules as follows: There exists a finite $F \subset \mathbb{Z}^d$ and $\Phi : \mathcal{A}^F \to \mathcal{B}$ such that $\forall x \in X : \phi(x)_v = \Phi(\sigma_{-v}(x)|_F)$. If ϕ is an endomorphism then we refer to it as a cellular automaton. A cellular automaton is said to be reversible if there exists a cellular automaton ϕ^{-1} such that $\phi \circ \phi^{-1} = \phi^{-1} \circ \phi = \mathrm{id}$. It is well known that reversibility is equivalent to bijectivity.

Throughout this article we use the following notation inspired by Turing machines. We denote by $\Sigma = \{0, \ldots, n-1\}$ the set of tape symbols and $Q = \{1, \ldots, k\}$ the set of states. We also use exclusively the symbols $n = |\Sigma|$ for the size of the alphabet and $k = |Q|$ for the number of states. Given a function $f : \Omega \to \prod_{i \in I} A_i$ we denote by f_i the projection of f to the i-th coordinate.

2   Two Models for Turing Machine Groups

In this section we define our generalized Turing machine model, and the group of Turing machines. In fact, we give two definitions for this group, one with a moving head and one with a moving tape as in [20]. We show that – except in the case of a trivial alphabet – these groups are isomorphic.[3] Furthermore, both can be defined both by local rules and 'dynamically', that is, in terms of continuity and the shift. In the moving tape model we characterize reversibility as preservation of the uniform measure. Finally we conclude this section by characterizing reversibility for classical Turing machines in our setting.

2.1   The Moving Head Model

Consider $Q = \{1, \ldots, k\}$ and let X_k be the subshift with alphabet $Q \cup \{0\}$ such that in each configuration the number of non-zero symbols is at most one.

$$X_k = \{x \in \{0, 1, \ldots, k\}^{\mathbb{Z}^d} \mid 0 \notin \{x_u, x_v\} \implies u = v\}.$$

In particular $X_0 = \{0^{\mathbb{Z}^d}\}$ and $i < j \implies X_i \subsetneq X_j$. Let also $\Sigma = \{0, \ldots, n - 1\}$ and $X_{n,k} = \Sigma^{\mathbb{Z}^d} \times X_k$. For the case $d = 1$, configurations in $X_{n,k}$ represent a bi-infinite tape filled with symbols in Σ possibly containing a head that has a state in Q. Note that there might be no head in a configuration.

Definition 2. *Given a function*

$$f : \Sigma^F \times Q \to \Sigma^{F'} \times Q \times \mathbb{Z}^d,$$

where F, F' are finite subsets of $\mathbb{Z}^d$, we can define a map $T_f : X_{n,k} \to X_{n,k}$ as follows: Let $(x, y) \in X_{n,k}$. If there is no $v \in \mathbb{Z}^d$ such that $y_v \neq 0$ then $T(x, y) = (x, y)$. Otherwise let $p = \sigma_{-v}(x)|_F$, $q = y_v \neq 0$ and $f(p, q) = (p', q', d)$. Then $T(x, y) = (\tilde{x}, \tilde{y})$ where:

$$\tilde{x}_t = \begin{cases} x_t & \text{if } t - v \notin F' \\ p'_{t-v} & \text{if } t - v \in F' \end{cases}, \qquad \tilde{y}_t = \begin{cases} q' & \text{if } t = v + d \\ 0 & \text{otherwise} \end{cases}$$

Such $T = T_f$ is called a (moving head) (d, n, k)-Turing machine, and f is its local rule. If there exists a (d, n, k)-Turing machine T^{-1} such that $T \circ T^{-1} = T^{-1} \circ T = \text{id}$, we say T is reversible.

---

[3] Note that the *dynamics* obtained from these two definitions are in fact quite different, as shown in [20, 21].

Note that $\sigma_{-v}(x)|_F$ is the F-shaped pattern 'at' v, but we do not write $x|_{F+v}$ because we want the pattern we read from x to have F as its domain.

This definition corresponds to classical Turing machines with the moving head model when $d = 1$, $F = F' = \{0\}$ and $f(x, q)_3 \in \{-1, 0, 1\}$ for all x, q. By possibly changing the local rule f, we can always choose $F = [-r_i, r_i]^d$ and $F' = [-r_o, r_o]^d$ for some $r_i, r_o \in \mathbb{N}$, without changing the Turing machine T_f it defines. The minimal such r_i is called the *in-radius* of T, and the minimal r_o is called the *out-radius* of T. We say the in-radius of a Turing machine is -1 if there is no dependence on input, that is, the neighborhood $[-r_i, r_i]$ can be replaced by the empty set. Since $\Sigma^F \times Q$ is finite, the third component of $f(p, q)$ takes only finitely many values $v \in \mathbb{Z}^d$. The maximum of $|v|$ for such v is called the *move-radius* of T. Finally, the maximum of all these three radii is called the *radius* of T. In this terminology, classical Turing machines are those with in- and out-radius 0, and move-radius 1.

Definition 3. *Define* TM($\mathbb{Z}^d, n, k$) *as the set of* (d, n, k)-*Turing machines and* RTM($\mathbb{Z}^d, n, k$) *the set of reversible* (d, n, k)-*Turing machines.*

In some parts of this article we just consider $d = 1$. In this case we simplify the notation and just write RTM(n, k) := RTM($\mathbb{Z}, n, k$).

Of course, we want TM($\mathbb{Z}^d, n, k$) to be a monoid and RTM($\mathbb{Z}^d, n, k$) a group under function composition. This is indeed the case, and one can prove this directly by constructing local rules for the inverse of a reversible Turing machine and composition of two Turing machines. However, it is much easier to extract this from the following characterization of Turing machines as a particular kind of cellular automaton.

For a subshift X, we denote by End(X) the monoid of endomorphisms of X and Aut(X) the group of automorphisms of X.

Proposition 1. *Let* n, k *be positive integers and* $Y = X_{n,0}$. *Then:*

$$\text{TM}(\mathbb{Z}^d, n, k) = \{\phi \in \text{End}(X_{n,k}) \mid \phi|_Y = \text{id}, \phi^{-1}(Y) = Y\}$$
$$\text{RTM}(\mathbb{Z}^d, n, k) = \{\phi \in \text{Aut}(X_{n,k}) \mid \phi|_Y = \text{id}\}$$

Corollary 1. *We have* $\phi \in$ RTM($\mathbb{Z}^d, n, k$) *if and only if* $\phi \in$ TM($\mathbb{Z}^d, n, k$) *and* ϕ *is bijective.*

Clearly, the conditions of Proposition 1 are preserved under function composition and inversion. Thus:

Corollary 2. *Under function composition,* (TM($\mathbb{Z}^d, n, k$), $\circ$) *is a submonoid of* End($X_{n,k}$) *and* (RTM($\mathbb{Z}^d, n, k$), $\circ$) *is a group.*

We usually omit the function composition symbol, and use the notations TM($\mathbb{Z}^d, n, k$) and RTM($\mathbb{Z}^d, n, k$) to refer to the corresponding monoids and groups.

2.2   The Moving Tape Model

It's also possible to consider the position of the Turing machine as fixed at 0, and move the tape instead, to obtain the moving tape Turing machine model. In [20], where Turing machines are studied as dynamical systems, the moving head model and moving tape model give non-conjugate dynamical systems. However, the abstract monoids defined by the two points of view turn out to be equal, and we obtain an equivalent definition of the group of Turing machines.

As in the previous section, we begin with a definition using local rules.

Definition 4. *Given a function* $f : \Sigma^F \times Q \to \Sigma^{F'} \times Q \times \mathbb{Z}^d$, *where* F, F' *are finite subsets of* $\mathbb{Z}^d$, *we can define a map* $T_f : \Sigma^{\mathbb{Z}^d} \times Q \to \Sigma^{\mathbb{Z}^d} \times Q$ *as follows: If* $f(x|_F, q) = (p, q', \boldsymbol{d})$, *then* $T_f(x, q) = (\sigma_{\boldsymbol{d}}(y), q')$ *where*

$$y_{\boldsymbol{u}} = \begin{cases} x_{\boldsymbol{u}}, \text{ if } \boldsymbol{u} \notin F' \\ p_{\boldsymbol{u}}, \text{ if } \boldsymbol{u} \in F', \end{cases}$$

is called the moving tape Turing machine *defined by* f.

These machines also have the following characterization with a slightly more dynamical feel to it. Say that x and y are *asymptotic*, and write $x \sim y$, if $d(\sigma_{\boldsymbol{v}}(x), \sigma_{\boldsymbol{v}}(y)) \to 0$ as $|\boldsymbol{v}| \to \infty$. We write $x \sim_m y$ if $x_{\boldsymbol{v}} = y_{\boldsymbol{v}}$ for all $|\boldsymbol{v}| \geq m$, and clearly $x \sim y \iff \exists m : x \sim_m y$.

Lemma 1. *Let* $T : \Sigma^{\mathbb{Z}^d} \times Q \to \Sigma^{\mathbb{Z}^d} \times Q$ *be a function. Then* T *is a moving tape Turing machine if and only if it is continuous, and for a continuous function* $s :$ $\Sigma^{\mathbb{Z}^d} \times Q \to \mathbb{Z}^d$ *and* $a \in \mathbb{N}$ *we have* $T(x, q)_1 \sim_a \sigma_{s(x,q)}(x)$ *for all* $(x, q) \in \Sigma^{\mathbb{Z}^d} \times Q$.

Note that in place of a we could allow a continuous $\mathbb{N}$-valued function of (x, q) – the definition obtained would be equivalent, as the a of the present definition can be taken as the maximum of such a function.

We call the function s in the definition of these machines the *shift indicator* of T, as it indicates how much the tape is shifted depending on the local configuration around 0. In the theory of orbit equivalence and topological full groups, the analogs of s are usually called *cocycles*. We also define in-, out- and move-radii of moving tape Turing machines similarly as in the moving head case.

We note that it is not enough that $T(x, q)_1 \sim \sigma_{s(x,q)}(x)$ for all $(x, q) \in$ $\Sigma^{\mathbb{Z}^d} \times Q$: Let $Q = \{1\}$ and consider the function $T : \Sigma^{\mathbb{Z}} \times Q \to \Sigma^{\mathbb{Z}} \times Q$ defined by $(T(x, 1)_1)_i = x_{-i}$ if $x_{[-|i|+1, |i|-1]} = 0^{2i-1}$ and $\{x_i, x_{-i}\} \neq \{0\}$, and $(T(x, 1)_1)_i = x_i$ otherwise. Clearly this map is continuous, and the constant-$\boldsymbol{0}$ map $s(x, q) = \boldsymbol{0}$ gives a shift-indicator for it. However, T is not defined by any local rule since it can modify the tape arbitrarily far from the origin.

As for moving head machines, it is easy to see (either by constructing local rules or by applying the dynamical definition) that the composition of two moving tape Turing machines is again a moving tape Turing machine. This allows us to proceed as before and define their monoid and group.

Definition 5. *We denote by* $\mathrm{TM}_{\mathrm{fix}}(\mathbb{Z}^d, n, k)$ *and* $\mathrm{RTM}_{\mathrm{fix}}(\mathbb{Z}^d, n, k)$ *the monoid of moving tape* (d, n, k)-*Turing machines and the group of reversible moving tape* (d, n, k)-*Turing machines respectively.*

Now, let us show that the moving head and moving tape models are equivalent. First, there is a natural epimorphism $\Psi : \mathrm{TM}(\mathbb{Z}^d, n, k) \to \mathrm{TM}_{\mathrm{fix}}(\mathbb{Z}^d, n, k)$. Namely, let $T \in \mathrm{TM}(\mathbb{Z}^d, n, k)$. We define $\Psi(T)$ as follows: Let $(x, q) \in \Sigma^{\mathbb{Z}^d} \times Q$. Letting y be the configuration such that $y_0 = q$ and 0 everywhere else and $T(x, y) = (x', y')$ such that $y'_v = q'$ we define $\Psi(T)(x, q) = (\sigma_{-v}(x'), q')$. This is clearly an epimorphism but it's not necessarily injective if $n = 1$. Indeed, we have that $\mathrm{RTM}_{\mathrm{fix}}(\mathbb{Z}^d, 1, k) \cong S_k$ and $\mathrm{TM}_{\mathrm{fix}}(\mathbb{Z}^d, 1, k)$ is isomorphic to the monoid of all functions from $\{1, \dots, k\}$ to itself while $\mathbb{Z}^d \leq \mathrm{RTM}(\mathbb{Z}^d, 1, k) \leq \mathrm{TM}(\mathbb{Z}^d, 1, k)$. Nevertheless, if $n \geq 2$ this mapping is an isomorphism.

Lemma 2. *If* $n \geq 2$ *then:*

$$\mathrm{TM}_{\mathrm{fix}}(\mathbb{Z}^d, n, k) \cong \mathrm{TM}(\mathbb{Z}^d, n, k)$$
$$\mathrm{RTM}_{\mathrm{fix}}(\mathbb{Z}^d, n, k) \cong \mathrm{RTM}(\mathbb{Z}^d, n, k).$$

The previous result means that besides the trivial case $n = 1$ where the tape plays no role, we can study the properties of these groups using any model.

2.3 The Uniform Measure and Reversibility

Consider the space $\Sigma^{\mathbb{Z}^d} \times Q$. We define a measure μ on $\mathcal{B}(\Sigma^{\mathbb{Z}^d} \times Q)$ as the product measure of the uniform Bernoulli measure and the uniform discrete measure. That is, if F is a finite subset of $\mathbb{Z}^d$ and $p \in \Sigma^F$, then:

$$\mu([p] \times \{q\}) = \frac{1}{kn^{|F|}}.$$

With this measure in hand we can prove the following:

Theorem 1. *Let* $T \in \mathrm{TM}_{\mathrm{fix}}(\mathbb{Z}^d, n, k)$. *Then the following are equivalent:*

1. *T is injective.*
2. *T is surjective.*
3. *$T \in \mathrm{RTM}_{\mathrm{fix}}(\mathbb{Z}^d, n, k)$.*
4. *T preserves the uniform measure* ($\mu(T^{-1}(A)) = \mu(A)$ *for all Borel sets* A).
5. *$\mu(T(A)) = \mu(A)$ for all Borel sets* A.

Remark 1. The proof is based on showing that every Turing machine is a *local homeomorphism* and preserves the measure of all large-radius cylinders in the forward sense. Note that preserving the measure of large-radius cylinders in the forward sense does not imply preserving the measure of all Borel sets, in general. For example, the machine which turns the symbol in $F = \{0\}$ to 0

without moving the head satisfies $\mu([p]) = \mu(T[p])$ for any $p \in \Sigma^S$ with $S \supset F$. But $\mu([]) = 1$ and $\mu(T([])) = \mu([0]) = 1/2$, where $[] = \{0, \ldots, n\}^{\mathbb{Z}}$ is the cylinder defined by the empty word.

Using the measure, one can define the average movement of a Turing machine.

Definition 6. *Let* $T \in \mathrm{TM}_{\mathrm{fix}}(\mathbb{Z}^d, n, k)$ *with shift indicator function* $s : \Sigma^{\mathbb{Z}^d} \times Q \to \mathbb{Z}^d$. *We define the* average movement $\alpha(T) \in \mathbb{Q}^d$ *as*

$$\alpha(T) := \mathbb{E}_\mu(s) = \int_{\Sigma^{\mathbb{Z}^d} \times Q} s(x, q) d\mu,$$

where μ *is the uniform measure defined in Subsect. 2.3. For* T *in* $\mathrm{TM}(\mathbb{Z}^d, n, k)$ *we define* α *as the application to its image under the canonical epimorphism* Ψ, *that is,* $\alpha(T) := \alpha(\Psi(T))$.

We remark that this integral is actually a finite sum over the cylinders $p \in \Sigma^F$. Nonetheless, its expression as an expected value allows us to show the following: If $T_1, T_2 \in \mathrm{RTM}(\mathbb{Z}^d, n, k)$ then $\alpha(T_1 \circ T_2) = \alpha(T_1) + \alpha(T_2)$. Indeed, as reversibility implies measure-preservation, we have that

$$\mathbb{E}_\mu(s_{T_1 \circ T_2}) = \mathbb{E}_\mu(s_{T_1} \circ T_2 + s_{T_2}) = \mathbb{E}_\mu(s_{T_1}) + \mathbb{E}_\mu(s_{T_2}).$$

This means that α defines an homomorphism from $\mathrm{RTM}(\mathbb{Z}^d, n, k)$ to $\mathbb{Q}^d$.

2.4 Classical Turing Machines

As discussed in the introduction, we say a one-dimensional Turing machine is *classical* if its in- and out-radii are 0, and its move-radius is 1. In this section, we characterize reversibility in classical Turing machines. If T_0 has in-, out- and move-radius 0, that is, T_0 only performs a permutation of the set of pairs $(s, q) \in \Sigma \times Q$ at the position of the head, then we say T_0 is a *state-symbol permutation*. If T_1 has in-radius -1, never modifies the tape, and only makes movements by vectors in $\{-1, 0, 1\}$, then T_1 is called a *state-dependent shift*.[4]

Theorem 2. *A classical Turing machine* T *is reversible if and only if it is of the form* $T_1 \circ T_0$ *where* T_0 *is a state-symbol permutation and* T_1 *is a state-dependent shift.*

It follows that the inverse of a reversible classical Turing machine is always of the form $T_0 \circ T_1$ where T_0 is a state-symbol permutation and T_1 is a state-dependent shift. In the terminology of Sect. 3, the theorem implies that all reversible classical Turing machines are elementary.

---

[4] Note that these machines are slightly different than the groups $\mathrm{SP}(\mathbb{Z}, n, k)$ and $\mathrm{Shift}(\mathbb{Z}, n, k)$ introduced in Sect. 3, as the permutations in $\mathrm{SP}(\mathbb{Z}, n, k)$ do not modify the tape, and moves in $\mathrm{Shift}(\mathbb{Z}; n, k)$ cannot depend on the state.

3   Properties of RTM and Interesting Subgroups

In this section we study some properties of RTM by studying the subgroups it contains. We introduce LP, the group of local permutations where the head does not move and RFA, the group of (reversible) finite-state automata which do not change the tape. These groups separately capture the dynamics of changing the tape and moving the head. We also define the group of oblivious Turing machines OB as an extension of LP where arbitrary tape-independent moves are allowed, and EL as the group of elementary Turing machines, which are compositions of finite-state automata and oblivious Turing machines.

First, we observe that $\alpha(\mathrm{RTM}(\mathbb{Z}^d, n, k))$ is not finitely generated, and thus:

Theorem 3. *For $n \geq 2$, the group $\mathrm{RTM}(\mathbb{Z}^d, n, k)$ is not finitely generated.*

Although α is not a homomorphism on $\mathrm{TM}(\mathbb{Z}^d, n, k)$, using Theorem 1, we obtain that $\mathrm{TM}(\mathbb{Z}^d, n, k)$ cannot be finitely generated either.

3.1   Local Permutations and Oblivious Turing Machines

For $\boldsymbol{v} \in \mathbb{Z}^d$, define the machine $T_{\boldsymbol{v}}$ which does not modifies the state or the tape, and moves the head by the vector $\boldsymbol{v}$ on each step. Denote the group of such machines by $\mathrm{Shift}(\mathbb{Z}^d, n, k)$. Clearly $\alpha : \mathrm{Shift}(\mathbb{Z}^d, n, k) \to \mathbb{Z}^d$ is a group isomorphism. Define also $\mathrm{SP}(\mathbb{Z}^d, n, k)$ as the *state-permutations*: Turing machines that never move and only permute their state as a function of the tape.

Definition 7. *We define the group $\mathrm{LP}(\mathbb{Z}^d, n, k)$ of local permutations as the subgroup of reversible (d, n, k)-Turing machines whose shift-indicator is the constant-$\boldsymbol{0}$ function. Define also $\mathrm{OB}(\mathbb{Z}^d, n, k) = \langle \mathrm{Shift}(\mathbb{Z}^d, n, k), \mathrm{LP}(\mathbb{Z}^d, n, k) \rangle$, the group of oblivious Turing machines.*

In other words, $\mathrm{LP}(\mathbb{Z}^d, n, k)$ is the group of reversible machines that do not move the head, and $\mathrm{OB}(\mathbb{Z}^d, n, k)$ is the group of reversible Turing machines whose head movement is independent of the tape contents. Note that in the definition of both groups, we allow changing the state as a function of the tape, and vice versa. Clearly $\mathrm{Shift}(\mathbb{Z}^d, n, k) \leq \mathrm{OB}(\mathbb{Z}^d, n, k)$ and $\mathrm{SP}(\mathbb{Z}^d, n, k) \leq \mathrm{LP}(\mathbb{Z}^d, n, k)$.

Proposition 2. *Let S_∞ be the group of permutations of $\mathbb{N}$ of finite support. Then for $n \geq 2$, $S_\infty \hookrightarrow \mathrm{LP}(\mathbb{Z}^d, n, k)$.*

In particular, $\mathrm{RTM}(\mathbb{Z}^d, n, k)$ is not residually finite. By Cayley's theorem, Proposition 2 also implies that $\mathrm{RTM}(\mathbb{Z}^d, n, k)$ contains all finite groups.

Proposition 3. *The group $\mathrm{OB}(\mathbb{Z}^d, n, k)$ is amenable.*

Write $H \wr G$ for the restricted wreath product.

Proposition 4. *If G is a finite group and $n \geq 2$, then $G \wr \mathbb{Z}^d \hookrightarrow \mathrm{OB}(\mathbb{Z}^d, n, k)$.*

The groups $G \wr \mathbb{Z}^d$ are sometimes called generalized lamplighter groups. In fact, $\mathrm{OB}(\mathbb{Z}^d, n, k)$ can in some sense be seen as a generalized generalized lamplighter group, since the subgroup of $\mathrm{OB}(\mathbb{Z}^d, n, k)$ generated by the local permutations $\mathrm{LP}(\mathbb{Z}^d, n, 1)$ with radius 0 and $\mathrm{Shift}(\mathbb{Z}^d, n, 1)$ is isomorphic to $A \cong S_n \wr \mathbb{Z}^d$.

Interestingly, just like the generalized lamplighter groups, we can show that the whole group $\mathrm{OB}(\mathbb{Z}^d, n, k)$ is finitely generated.

Theorem 4. $\mathrm{OB}(\mathbb{Z}^d, n, k)$ *is finitely generated.*

3.2   Finite-State Automata

Definition 8. *We define the* reversible finite-state automata $\mathrm{RFA}(\mathbb{Z}^d, n, k)$ *as the group of reversible* (d, n, k)*-Turing machines that do not change the tape. That is, the local rules are of the form* $f(p, q) = (p, q', z)$ *for all entries* $p \in \Sigma^F, q \in Q$.

This group is orthogonal to $\mathrm{OB}(\mathbb{Z}^d, n, k)$ in the following sense:

Proposition 5

$$\mathrm{RFA}(\mathbb{Z}^d, n, k) \cap \mathrm{LP}(\mathbb{Z}^d, n, k) = \mathrm{SP}(\mathbb{Z}^d, n, k)$$
$$\mathrm{RFA}(\mathbb{Z}^d, n, k) \cap \mathrm{OB}(\mathbb{Z}^d, n, k) = \langle \mathrm{SP}(\mathbb{Z}^d, n, k), \mathrm{Shift}(\mathbb{Z}^d, n, k) \rangle$$

As usual, the case $n = 1$ is not particularly interesting, and we have that $\mathrm{RFA}(\mathbb{Z}^d, 1, k) \cong \mathrm{RTM}(\mathbb{Z}^d, 1, k)$. In the general case the group is more complex.

We now prove that the $\mathrm{RFA}(\mathbb{Z}^d, n, k)$-groups are non-amenable. In [10], a similar idea is used to prove that there exists a minimal $\mathbb{Z}^2$-subshift whose topological full group is not amenable.

Proposition 6. *Let* $n \geq 2$. *For all* $m \in \mathbb{N}$ *we have that:*

$$\underbrace{\mathbb{Z}/2\mathbb{Z} * \cdots * \mathbb{Z}/2\mathbb{Z}}_{m \text{ times}} \hookrightarrow \mathrm{RFA}(\mathbb{Z}^d, n, k)$$

Corollary 3. *For* $n \geq 2$, $\mathrm{RFA}(\mathbb{Z}^d, n, k)$ *and* $\mathrm{RTM}(\mathbb{Z}^d, n, k)$ *contain the free group on two elements. In particular, they are not amenable.*

By standard marker constructions, one can also embed all finite groups and finitely generated abelian groups in $\mathrm{RFA}(\mathbb{Z}^d, n, k)$ – however, this group is residually finite, and thus does not contain S_∞ or $(\mathbb{Q}, +)$.

Proposition 7. *Let* $n \geq 2$ *and* G *be any finite group or a finitely generated abelian group. Then* $G \leq \mathrm{RFA}(\mathbb{Z}^d, n, k)$.

Theorem 5. *Let* $n \geq 2, k \geq 1, d \geq 1$. *Then the group* $\mathrm{RFA}(\mathbb{Z}^d, n, k)$ *is residually finite and is not finitely generated.*

The proof of this theorem is based on studying the action of the group on finite subshifts where heads are occur periodically. Non-finitely generatedness is obtained by looking at signs of permutations of the finitely many orbits, to obtain the *sign homomorphism* to an infinitely generated abelian group.

3.3 Elementary Turing Machines and the LEF Property of RTM

Definition 9. *We define the group of elementary Turing machines* $\mathrm{EL}(\mathbb{Z}^d, n, k) := \langle \mathrm{RFA}(\mathbb{Z}^d, n, k), \mathrm{LP}(\mathbb{Z}^d, n, k) \rangle$. *That is, the group generated by machines which only change the tape or move the head.*

Proposition 8. *Let* $\mathbb{Q}_p = \frac{1}{p}\mathbb{Z}$. *Then* $\alpha(\mathrm{RFA}(\mathbb{Z}^d, n, k)) = \alpha(\mathrm{EL}(\mathbb{Z}^d, n, k)) = \mathbb{Q}_k^d$. *In particular,* $\mathrm{EL}(\mathbb{Z}^d, n, k) \subsetneq \mathrm{RTM}(\mathbb{Z}^d, n, k)$.

We do not know whether $\alpha(T) \in \mathbb{Z}^d$ implies $T \in \mathrm{EL}(\mathbb{Z}^d, n, 1)$, nor whether $\mathrm{EL}(\mathbb{Z}^d, n, k)$ is finitely generated – the sign homomorphism we use in the proof of finitely-generatedness of the group of finite-state automata does not extend to it.

By the results of this section, the group $\mathrm{RTM}(\mathbb{Z}^d, n, k)$ is neither amenable nor residually finite. By adapting the proof of Theorem 5, one can show that it is locally embeddable in finite groups. See [29–31] for the definitions.

Theorem 6. *The group* $\mathrm{RTM}(\mathbb{Z}^d, n, k)$ *is LEF, and thus sofic, for all* n, k, d.

4 Computability Aspects

4.1 Basic Decidability Results

First, we observe that basic management of local rules is decidable. Note that these results hold, and are easy to prove, even in higher dimensions.

Lemma 3. *Given two local rules* $f, g : \Sigma^F \times Q \to \Sigma^F \times Q \times \mathbb{Z}^d$,

- *it is decidable whether* $T_f = T_g$,
- *we can effectively compute a local rule for* $T_f \circ T_g$,
- *it is decidable whether* T_f *is reversible, and*
- *we can effectively compute a local rule for* T_f^{-1} *when* T_f *is reversible.*

A group is called *recursively presented* if one can algorithmically enumerate its elements, and all identities that hold between them. If one can furthermore decide whether a given identity holds in the group (equivalently, whether a given element is equal to the identity element), we say the group has a *decidable word problem*. The above lemma is the algorithmic content of the following proposition:

Proposition 9. *The groups* $\mathrm{TM}(\mathbb{Z}^d, n, k)$ *and* $\mathrm{RTM}(\mathbb{Z}^d, n, k)$ *are recursively presented and have decidable word problems in the standard presentations.*

4.2 The Torsion Problem

The *torsion problem* of a recursively presented group G is the set of presentations of elements $g \in G$ such that $g^n = 1_G$ for some $n \geq 1$. Torsion elements are recursively enumerable when the group G is recursively presented, but the torsion problem need not be decidable even when G has decidable word problem.

In the case of $\mathrm{RTM}(\mathbb{Z}^d, n, k)$ the torsion problem is undecidable for $n \geq 2$. This result was shown by Kari and Ollinger in [19] using a reduction from the mortality problem which they also prove to be undecidable.

The question becomes quite interesting if we consider the subgroup $\mathrm{RFA}(\mathbb{Z}^d, n, k)$ for $n \geq 2$, as then the decidability of the torsion problem is dimension-sensitive.

Theorem 7. *The torsion problem of* $\mathrm{RFA}(\mathbb{Z}, n, k)$ *is decidable.*

Theorem 8. *For all* $n \geq 2, k \geq 1, d \geq 2$, *there is a finitely generated subgroup of* $\mathrm{RFA}(\mathbb{Z}^d, n, k)$ *whose torsion problem is undecidable.*

Acknowledgements. The third author was supported by FONDECYT grant 3150552.

References

1. Ollinger, N., Gajardo, A., Torres-Avilés, R.: The transitivity problem of turing machines (2015)
2. Aaronson, S., Grier, D., Schaeffer, L.: The classification of reversible bit operations. ArXiv e-prints, April 2015
3. Aubrun, N., Barbieri, S., Sablik, M.: A notion of effectiveness for subshifts on finitely generated groups. ArXiv e-prints, December 2014
4. Aubrun, N., Sablik, M.: Simulation of effective subshifts by two-dimensional subshifts of finite type. Acta Applicandae Math. **126**(1), 35–63 (2013)
5. Belk, J., Bleak, C.: Some undecidability results for asynchronous transducers and the Brin-Thompson group 2V. ArXiv e-prints, May 2014
6. Belk, J., Matucci, F.: Conjugacy and dynamics in Thompson's groups. Geom. Dedicata **169**(1), 239–261 (2014)
7. Ceccherini-Silberstein, T., Coornaert, M.: Cellular Automata and Groups. Springer Monographs in Mathematics. Springer, Heidelberg (2010)
8. Czeizler, E., Kari, J.: A tight linear bound on the synchronization delay of bijective automata. Theor. Comput. Sci. **380**(12), 23–36 (2007). Automata, Languages and Programming
9. Durand, B., Romashchenko, A., Shen, A.: Effective closed subshifts in 1D can be implemented in 2D. In: Blass, A., Dershowitz, N., Reisig, W. (eds.) Fields of Logic and Computation. LNCS, vol. 6300, pp. 208–226. Springer, Heidelberg (2010)
10. Elek, G., Monod, N.: On the topological full group of a minimal Cantor Z^2-system. ArXiv e-prints, December 2012
11. Gajardo, A., Guillon, P.: Zigzags in turing machines. In: Ablayev, F., Mayr, E.W. (eds.) CSR 2010. LNCS, vol. 6072, pp. 109–119. Springer, Heidelberg (2010)
12. Gajardo, A., Mazoyer, J.: One head machines from a symbolic approach. Theor. Comput. Sci. **370**(13), 34–47 (2007)
13. Giordano, T., Putnam, I., Skau, C.: Full groups of cantor minimal systems. Isr. J. Math. **111**(1), 285–320 (1999)
14. Grigorchuk, R., Medynets, K.: On algebraic properties of topological full groups. ArXiv e-prints, May 2011
15. Kůrka, P., Delvenne, J., Blondel, V.: Decidability and universality in symbolic dynamical systems. Fund. Inform. **74**(4), 463–490 (2006)

16. Jeandel, E.: Computability of the entropy of one-tape Turing machines. arXiv preprint (2013). arXiv:1302.1170
17. Juschenko, K., Monod, N.: Cantor systems, piecewise translations and simple amenable groups. Ann. Math. (2012)
18. Kari, J.: Representation of reversible cellular automata with block permutations. Theor. Comput. Syst. **29**, 47–61 (1996). doi:10.1007/BF01201813
19. Kari, J., Ollinger, N.: Periodicity and immortality in reversible computing. In: Ochmański, E., Tyszkiewicz, J. (eds.) MFCS 2008. LNCS, vol. 5162, pp. 419–430. Springer, Heidelberg (2008)
20. Kůrka, P.: On topological dynamics of turing machines. Theor. Comput. Sci. **174**(1–2), 203–216 (1997)
21. Kůrka, P.: Erratum to: entropy of turing machines with moving head. Theor. Comput. Sci. **411**(31–33), 2999–3000 (2010)
22. Lind, D., Marcus, B.: An Introduction to Symbolic Dynamics and Coding. Cambridge University Press, Cambridge (1995)
23. Lind, D., Boyle, M., Rudolph, D.: The automorphism group of a shift of finite type. Trans. Am. Math. Soc. **306**(1), 71–114 (1988)
24. Matui, H.: Topological full groups of one-sided shifts of finite type. ArXiv e-prints, October 2012
25. Salo, V., Schraudner, M.: in preparation
26. Salo, V., Törmä, I.: Group-walking automata. In: Kari, J. (ed.) AUTOMATA 2015. LNCS, vol. 9099, pp. 224–237. Springer, Heidelberg (2015)
27. Salo, V., Törmä, I.: Plane-walking automata. In: Isokawa, T., Imai, K., Matsui, N., Peper, F., Umeo, H. (eds.) AUTOMATA 2014. LNCS, vol. 8996, pp. 135–148. Springer, Heidelberg (2015)
28. Cassaigne, J., Blondel, V., Nichitiu, C.: On the presence of periodic configurations in turing machines and in counter machines. Theor. Comput. Sci. **289**(1), 573–590 (2002)
29. Vershik, A., Gordon, E.: Groups that are locally embeddable in the class of finite groups. Algebra i Analiz **9**(1), 71–97 (1997)
30. Weiss, B.: Sofic groups and dynamical systems. Sankhyā Indian J. Stat. Ser. A **62**, 350–359 (2000)
31. Ziman, M., et al.: On finite approximations of groups and algebras. Ill. J. Math. **46**(3), 837–839 (2002)

Exact Discretization of 3-Speed Rational Signal Machines into Cellular Automata

Tom Besson$^{(\boxtimes)}$ and Jérôme Durand-Lose

University of Orléans, INSA Centre Val de Loire, LIFO EA 4022, Orléans, France
`tom.besson@univ-orleans.fr`

Abstract. Cellular Automata (CA) operate in discrete time and space whereas Signal Machines (SM) have been developed as a continuous idealization of CA capturing the key concept of signals/particles and collisions. Inside a Euclidean space, dimensionless signals move freely; collisions are instantaneous. Today's issue is the automatic generation of a CA *mimicking* a given SM. On the one hand, many ad hoc manual conversions exist. On the other hand, some irrational or 4+-speed SM exhibit Zeno-like behaviors/space-time compression or rely on information being locally unbounded, both being incompatible with CA. This article provides a solution to automatically generate an exactly mimicking CA for a restricted class of SM: the ones that uses only three rational speeds, and rational initial positions. In these SM, signals are always contained inside a regular mesh. The discretization brings forth the corresponding discrete mesh. The simulation is valid on any infinite run and preserves the relative position of collisions.

Keywords: Abstract geometrical computation · Automatic discretization · Cellular automata · Signal machines · Unconventional models of computation

1 Introduction

Cellular automata (CA) are massively synchronous, uniform and local discrete dynamical systems. Since their introduction by J. Von Neumann in the forties [17], a lot of research, from dynamics to algorithmic, have been made. However, a recurrent problem is to create CA for a specific purpose, exhibiting a given behavior. Dedicated CA are crafted by experts, in the same way that programming a computer used to be reserved to specialists. As of now, coding became easier and easier with more and more abstract languages. What would these be such for CA?

The model chosen to serve this purpose is the signal machines (SM). Signal machines, originally thought as an idealization of CA, focus on continuous counterparts of (discrete) signals and collisions in CA. Signals and collisions are indeed key concepts of CA. They provide a lot of insight on the fabric of CA and

© IFIP International Federation for Information Processing 2016
Published by Springer International Publishing Switzerland 2016. All Rights Reserved
M. Cook and T. Neary (Eds.): AUTOMATA 2016, LNCS 9664, pp. 63–76, 2016.
DOI: 10.1007/978-3-319-39300-1_6

are the foundation of collision computing [1]. Some of their first usages stand in the generation of prime numbers by Fischer [13] and Goto's solution to the Firing Squad Synchronization Problem (FSSP) [14]. Signals are commonly used: to solve the FSSP [20], to compute with rule 110 [4] or with only four states [18], to understand one CA [5] and so on. In [19] one important question is the automatic positioning of discrete signals. Discrete signals are studied in [16].

In those references, as in many others, signals are represented by Euclidean lines to explain and reason as can be seen in Fischer's (Fig. 1(a)) and in Goto's (Fig. 1(b)) and Yunès's (Fig. 1(c)). Those constructions are understood in the continuum before being implemented in a discrete setting. Resulting states and transition functions are often left out because they are not used in the proof of constructions, obnoxious to establish, and cumbersome to read.

It is possible to *simulate* a CA with a SM [11]. However, the other way round is done only on particular ad hoc cases, often leaving the technical details out. The present paper provides an automatic conversion for a sub-class of signal machines: 3-speed rational. After a normalization phase, the states and the transition function are generated as well as initial configurations.

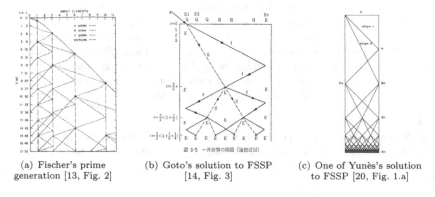

(a) Fischer's prime generation [13, Fig. 2]

(b) Goto's solution to FSSP [14, Fig. 3]

(c) One of Yunès's solution to FSSP [20, Fig. 1.a]

Fig. 1. Examples of signal use in designing cellular automata.

Signal machines are the formal tool for thinking about CA in the continuum and are inspired by the idealisation of discrete signals. They form an autonomous dynamical model where *signals* are dimensionless points moving with constant speed. They are completely described by their positions and natures called *metasignals*. Starting with a finite number of signals on the real axis, when two or more meet, they are destroyed and new signals are emitted. A set of *collision rules* defines which signals are emitted according to the colliding meta-signals. The dynamics is represented on a space-time diagram where signals appear as segments as in Fig. 2(b). In previous articles, one of the authors researched on the possibilities of the model, from its computational power [9,11], to its ability to simulate other models (like the BSS model [3] in [10]).

In signal machines, collisions are discrete steps related by signals: a collision is right before another if a signal generated by the first one ends in the second

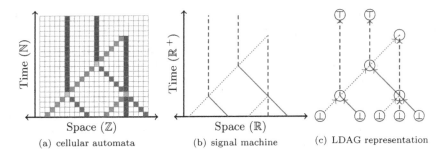

(a) cellular automata   (b) signal machine   (c) LDAG representation

Fig. 2. Identification of runs through the labeled DAG representation.

one. The dynamics of a run of a SM lies entirely in this *causal order* which can be represented as a Labeled Directed Acyclic Graph (LDAG). Figure 2 provides a LDAG identification of discrete and continuous space-time diagrams.

The present paper considers 3-speed rational SM: only three possible speeds are available and speeds as well as positions of signals in initial configurations are rational numbers. The behavior of these machines is limited since signals are trapped inside a regular, periodic-like *mesh* [2,12] which is essentially discrete (ensuring the discretization).

The existence of such meshes is not guaranteed with four or more speeds or irrationality. Moreover, if discretization is always locally possible it is not at a large scale nor for an infinite computation. Zeno-like behaviors/space-time compression known as *accumulation* (which are highly unpredictable [8]) leads to an infinite number of "objects" in bounded portion of a space-time diagram as depicted in Fig. 3(a) and (b).

An accumulation-less space-time diagram may not be discretizable into a CA as shown the one in Fig. 3(c). On the left, one solid signal is sent after $0, 1, 2, 3 \ldots$ dotted ones. This means that any number can be "stored" in-between

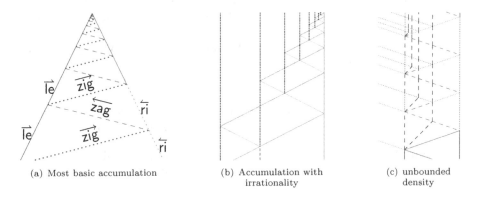

(a) Most basic accumulation   (b) Accumulation with irrationality   (c) unbounded density

Fig. 3. Non discretizable space-time diagrams.

the solid vertical signals. Discretization would lead to a finite number of cells and a bounded "storage capacity".

Outside the case covered in the present article, discretization is only possible on special cases designed for with an extra piece of information (whatever it may be) ensuring the process to work.

The automatic conversion starts with a normalization to get simple speeds. Then meta-signals and collision rules are turned into states and transitions. The translation of configurations follows the same patterns: normalization into integer positions then conversion. Special care is taken so that the discrete signals evolve inside a discrete mesh. The correctness of the process comes from the "preservation" of it and the LDAG inside it.

In Sect. 2, the SM and CA models are defined as well as SM-meshes and CA-signals. Section 3 focuses on dynamics and simulation as well as normalization of SM. Section 4 describes the discretization: generated states and transitions and initial configurations. Section 5 deals with the correctness of the construction through CA-meshes and LDAG. Section 6 gathers concluding remarks.

2   Definitions and Properties

A signal machine regroups the definitions of its meta-signals and their dynamics: constant speed outside of collisions and rewriting rules at collisions.

Definition 1. A *signal machine* (SM), $\mathfrak{A}$, is a triplet (M, S, R) such that: M is a finite set of *meta-signals*; $S : M \rightarrow \mathbb{R}$ is the *speed function* (each meta-signal has a constant speed); and R is a finite set of *collision rules* written $\rho = \rho^- \rightarrow \rho^+$ where ρ^- and ρ^+ are sets of meta-signals of distinct speeds. Each ρ^- must have at least two meta-signals. R is deterministic: $\rho \neq \rho'$ implies $\rho^- \neq \rho'$.

If no collision rule is defined for a set of meta-signals, the collision is *blank*: the same meta-signals are output ($\rho^+ = \rho^-$).

A *($\mathfrak{A}$-)configuration*, c, is a mapping from the real line to either a meta-signal, a collision rule or the value $\oslash$ indicating that there is nothing there. There are finitely many non-$\oslash$ locations.

If there is a signal of speed s at x, then after a duration Δt its position is $x + s \cdot \Delta t$, unless it enters a collision before. At a collision, all incoming signals are immediately replaced according to collision rules by outgoing signals in the following configurations.

A *space-time diagram* is the collection of consecutive configurations forming a two dimensional picture (time is always elapsing upwards as in Fig. 2).

Definition 2. A signal machine is *3-speed rational* (3S-$\mathbb{Q}$) if: only three speeds are available for the meta-signals and all speeds are rational numbers as well as any non-$\oslash$ position in any initial configuration.

These machines have their signals trapped inside a mesh made of the union of half-lines following the three speeds like the one in Fig. 4.

Definition 3. Let p, q and n be positive integers, p and q relatively prime, the (p, q, n)-*(SM-)mesh* corresponds to the union of the following half-lines of $\mathbb{R} \times \mathbb{R}^+$:

- $\mathfrak{V}_v : 0 \leq t$ and $x = v/(p+q)$ where $v \in \{0, 1, 2, \ldots, n(p+q)\}$,
- $\mathfrak{L}_l : x \leq n$ and $x = (l - t)/q$ where $l \in \mathbb{N}$, and
- $\mathfrak{R}_r : 0 \leq x$ and $x = (t - r)/p$ where $r \in \{(-n.p), \ldots, -1, 0, 1, \ldots\}$.

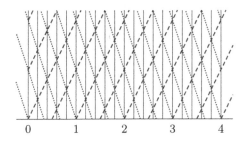

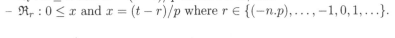

$$0 \qquad 1 \qquad 2 \qquad 3 \qquad 4$$

Fig. 4. The $(2, 3, 4)$-mesh.

Properties 1 ([2, **Lemma 1**] and [12, **Lemma 1**]). *In the space-time diagram generated from a $3s - \mathbb{Q}$ signal machine with speeds $-\frac{1}{q}$, 0 and $\frac{1}{p}$, p and q relatively prime, on an initial configuration where non-$\oslash$ values are in $\{0, 1, \ldots, n\}$, all the non-$\oslash$ positions belong to the (p, q, n)-(SM-)mesh.*

In the (p, q, n)-mesh, the *(SM-)encounter* of coordinates (v, r), $\mathfrak{e}_v^r$, is the intersection of $\mathfrak{V}_v$, $\mathfrak{R}_r$ and $\mathfrak{L}_{p+q}$; that is $\left(\frac{v}{p+q}, r + \frac{p}{p+q}v\right)$. Encounter $\mathfrak{e}_v^{r+1}$ directly depends on signals coming from $\mathfrak{e}_{v-1}^{r+1}$, $\mathfrak{e}_v^r$, and $\mathfrak{e}_{v+1}^{r-1}$ (if they exist) or the initial configuration. The integers v and r are such that $0 \leq v \leq n(p+q)$ and $-n.q \leq r$. Encounters ordered according to dependencies form a well founded order used for inductive proofs.

Definition 4. A (1-dimensional radius-1) *cellular automaton* (CA) is a triplet $(Q, f, \#)$ such that: Q is a finite set of *states*, $f : Q^3 \to Q$ is the local *transition function*, and $\#$ is a special state such that $f(\#, \#, \#) = \#$ (the *quiescent* state). A *configuration* maps cells to states, i.e. it is an element of $Q^{\mathbb{Z}}$. In the evolution from a configuration c, the *site*, c_x^t, is the cell $x(\in X)$ at time-step t ($\in \mathbb{N}$). It is computed by $c_x^{t+1} = f(c_{x-1}^t, c_x^t, c_{x+1}^t)$.

Definition 5. Let $W = M \cup R$ be the set of meta-signals and collision rules of a signal machine $\mathfrak{A}$ and Q be the set of states of a CA $\mathcal{A}$. A $\mathfrak{A}$-$\mathcal{A}$ representation relation $\mathcal{R}$ relates W and Q with the intended meaning that $\lambda \mathcal{R} q$ iff λ is *represented by* q. A state representing a collision rule does not represent anything else. A meta-signal can be represented by many states and a state can represent more than one meta-signal. A state representing nothing corresponds to $\oslash$.

Definition 6. For any meta-signal μ, a μ-CA-signal is defined by (x, a, b, φ) where: $x \in \mathbb{Z}$ is the *base position*, $a, b \in \mathbb{N}$ with $a \leq b$ (b can be $+\infty$) are the *birth* and *death* dates, and $\varphi \in [0, 1)$ is the *phase*. It corresponds to the set of sites:

- $\{ (x + \lfloor \varphi + (t - a).S(\mu) \rfloor, t) \mid t \in [a, b] \}$, if $0 < S(\mu)$ (rightward signal),
- $\{ (x - \lfloor \varphi - (t - a).S(\mu) \rfloor, t) \mid t \in [a, b] \}$, if $S(\mu) < 0$ (leftward signal), or
- $\{ (x, t) \mid t \in [a, b] \}$, otherwise (stationary signal).

It must represent $\mu : \mu \mathcal{R} c_x^t$ for all (x, t) in the μ-CA-Signal. Moreover it should be maximal (not extendable as a μ-CA-Signal). Its *speed* is the one of μ.

After the normalization, the non-stationary speeds are of the form $\pm 1/d$. Any non-stationary CA-signal has a simple periodic dynamics: it moves one cell on the side every d iterations. Figure 5 provides examples of CA-signals.

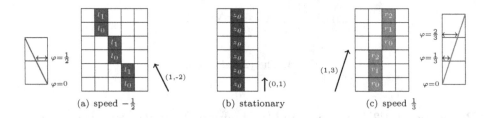

(a) speed $-\frac{1}{2}$ (b) stationary   (c) speed $\frac{1}{3}$

Fig. 5. Examples of CA-signals and phases.

For rightward signals, the phase φ is the distance between the bottom left corner of the cell and the place where the (continuous) signal enters the cell at the bottom. If the signal is leftward; the phase is the distance to the lower right corner of the cell. Phases are illustrated in Fig. 5.

3   Dynamics, Simulation and Normalization

The *causal order* of collisions in a space-time diagram can be represented as a Labeled Directed Acyclic Graph (LDAG) as illustrated in Fig. 2(c). If the LDAG's of two runs are identical, then the runs are *dynamic-wise* identical.

Formally, a $\mathfrak{A}$-*DAG* is a LDAG where edges (resp. vertices) are labeled with elements of M (resp. $R \cup \{\perp, \top\}$). Bottom (resp. top) leaves are exactly the vertices labeled with $\perp$ (resp. $\top$). The label $\perp$ is used for signals present in the initial configuration. The label $\top$ is used for never-ending signals. The LDAG generated from an initial configuration c is denoted $\hat{c}$.

A signal machine $\mathfrak{B}$ *simulates* a SM $\mathfrak{A}$ if there exist a *conversion function* $\zeta : \mathcal{C}_\mathfrak{A} \to \mathcal{C}_\mathfrak{B}$ and a *relabeling function* $\psi : M_\mathfrak{B} \cup R_\mathfrak{B} \to M_\mathfrak{A} \cup R_\mathfrak{A}$ such that: $\forall c \in \mathcal{C}_\mathfrak{A}, \ \hat{c} = \psi(\widehat{\zeta(c)})$ where ψ is canonically extended to LDAG as relabeling each edge and vertex.

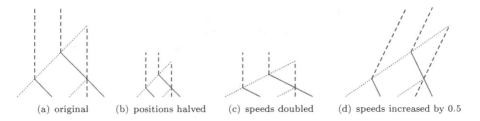

(a) original   (b) positions halved   (c) speeds doubled   (d) speeds increased by 0.5

Fig. 6. Examples of linear transformations.

Linearly (with positive coefficients) changing the speeds of the meta-signals or the initial positions does not affect the dynamics [7, Chap. 5]. Examples of such transformations are provided in Fig. 6.

A 3-speed rational signal machine is normalized with a positive linear operation so that its speeds are $-\frac{1}{q}, 0, \frac{1}{p}$ with p and q relatively prime. From now on, p and q are always used to refer to these denominators. Similarly, the positions in initial configurations are lifted onto $\mathbb{N}$.

Let $\mathcal{R}$ be a $\mathfrak{A}$-$\mathcal{A}$ representation relation, and c a $\mathcal{A}$-configuration. The LDAG $\widehat{c}^{\mathcal{R}}$, as illustrated in Fig. 2(c), is formed as follows. The edges correspond to all the CA-signals in the space-time diagram. The vertices correspond to all the sites that represent a collision rule plus one $\perp$ vertex for each CA-signal present in the initial configurations and one $\top$ vertex for each infinite CA-signal. The edges are in-incident to vertices if the vector from the topmost site of the signal to the vertex site is $(-1, 1)$, $(0, 1)$ or $(1, 1)$ (or an infinite CA-signal and the dedicated $\top$ vertex). The edges are out-incident to vertices if the vector from the bottom-most site of the signal to the vertex site is $(-1, -1)$, $(0, -1)$ or $(1, -1)$ (or present in the initial configuration and the dedicated $\perp$ vertex). The labels correspond to μ for μ-CA-signal and the (unique) collision rule represented by the vertex. If the LDAG is ill formed or non-unique, then $\widehat{c}^{\mathcal{R}}$ is undefined.

A cellular automaton $\mathcal{A}$ *simulates* a SM $\mathfrak{A}$ if there exists a *conversion function* $\zeta : \mathcal{C}_{\mathfrak{A}} \rightarrow \mathcal{C}_{\mathcal{A}}$ and a $\mathfrak{A}$-$\mathcal{A}$ *representation relation* $\mathcal{R}$ such that: $\forall c \in \mathcal{C}_{\mathfrak{A}}, \widehat{c} = \widehat{\zeta(c)}^{\mathcal{R}}$.

Dynamic-wise simulation preserves collision interactions but discard anything relevant to positions. Nothing prevents the resulting space-time diagrams to be "bent" (as in [6]) and one might want to preserve the geometry up to some linear operator (like *grouping* for CA in [15]).

To cope with this, we say that a simulation is *geometry-preserving* if the positions of the CA-collisions can be computed from the one of the corresponding SM-collision with a linear function (plus some integral rounding). The construction presented here is geometry-preserving.

4   Formal Discretization

The notation ⊟ defines the CA-transition $f(a, b, c) \rightarrow d$. The quiescent state # is left blank in the pictures. Any non-specified transition results in #. It is understood that r and r' are any rightward meta-signals, l and l' are any leftward ones and z and z' are any stationary ones.

The following states and transitions ensure the movement of isolated signals as illustrated in Fig. 5:

- for each (rightward) r: states r_k with $0 \leq k < p$ and

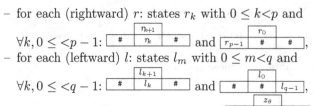

$\forall k, 0 \leq < p - 1$: [# | r_k $\overset{r_{k+1}}{}$ | #] and [r_{p-1} | # $\overset{r_0}{}$ | #],

- for each (leftward) l: states l_m with $0 \leq m < q$ and

$\forall k, 0 \leq < q - 1$: [# | l_k $\overset{l_{k+1}}{}$ | #] and [# | # $\overset{l_0}{}$ | l_{q-1}],

- for each (stationary) z: state z_θ and [# | z_θ $\overset{z_\theta}{}$ | #]. The subscript indicates a specific phase $\theta = \frac{q}{p+q}$ used to position exactly the collision.

When signals are moving closer as illustrated in Fig. 7, the following transitions are defined:

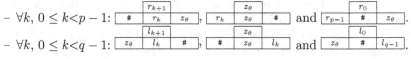

- $\forall k, 0 \leq k < p - 1$: [# | r_k $\overset{r_{k+1}}{}$ | z_θ], [r_k | z_θ $\overset{z_\theta}{}$ | #] and [r_{p-1} | # $\overset{r_0}{}$ | z_θ].
- $\forall k, 0 \leq k < q - 1$: [z_θ | l_k $\overset{l_{k+1}}{}$ | #], [# | z_θ $\overset{z_\theta}{}$ | l_k] and [z_θ | # $\overset{l_0}{}$ | l_{q-1}].

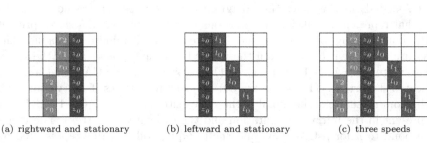

(a) rightward and stationary   (b) leftward and stationary   (c) three speeds

Fig. 7. Examples of 𝔄-CA-Signals closing on each other.

Let $\rho : \{r, z, l\} \to \{l', z', r'\}$ be any collision rule (any single r, z, l could be missing and any r', z', l' could be missing). When signals enter the same cell, the following phase hypothesis is assumed: non-stationary signals start with phase 0 and stationary ones with phase θ. The continuous signals are considered to locate precisely the collision. They correspond to the lines $y = px$ (rightward), $y = -qx + q$ (leftward) and $x = \theta$ (stationary). Resolving this system gives: $x = \frac{q}{p+q}$ and $y = \frac{p \cdot q}{p+q}$. The collisions always happen inside a cell because $\frac{p \cdot q}{p+q} \notin \mathbb{Z}$ (since p and q are relatively prime). The delay from getting in the same cell to the actual collision is denoted κ_0.

The states ρ_k represent the steps of the collision when two or more signals are inside the same cell: ρ_k means $(r_k : z_\theta : l_k)$ when $k < \kappa_0$ and $(r'_k : z'_\theta : l'_k)$ when $\kappa_0 < k$. Let κ_1 be the last time-step at which all the out-signals are inside the same cell and κ_2 the last time-step at which two out-signals are inside the same cell. Their values are given on Table 1 and illustrated in Fig. 8.

Table 1. Main steps of a collision.

$$\kappa_0 = \frac{p \cdot q}{p + q}$$

Emitted signals	z' and r'	l' and z'	l' and r'	l' and z' and r'
κ_1	$p - 1$	$q - 1$		$\min(p, q) - 1$
κ_2	/	/	/	$\max(p, q) - 1$

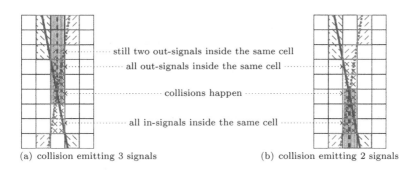

still two out-signals inside the same cell

all out-signals inside the same cell

collisions happen

all in-signals inside the same cell

(a) collision emitting 3 signals   (b) collision emitting 2 signals

Fig. 8. Illustration of κ_0, κ_1 and κ_2 with $p = 6$ and $q = 5$.

The signals enter the same cell (under the hypothesis) with $\boxed{\#\mid z_\theta\mid l_{q-1}}$ $(\rho_0$ above $z_\theta)$, $\boxed{r_{p-1}\mid z_\theta\mid \#}$ $(\rho_0$ above $z_\theta)$, $\boxed{r_{p-1}\mid \#\mid l_{q-1}}$ $(\rho_0$ above $\#)$, or $\boxed{r_{p-1}\mid z_\theta\mid l_{q-1}}$ $(\rho_0$ above $z_\theta)$. Then, they are getting closer until the exact collision: $\forall k,\ 0 \le k < \kappa_0$, $\boxed{\#\mid \rho_k\mid \#}$ $(\rho_{k+1}$ above $\rho_k)$.

If the collision emits no signal, $\boxed{\#\mid \rho_{\kappa_0}\mid \#}$ $(\#$ above $\rho_{\kappa_0})$ is added.

If the collision emits one signal as illustrated in Fig. 9(a), add:

- if $\rho^+ = \{r'\}$: $\boxed{\rho_{\kappa_0}\mid \#\mid \#}$ $(r'_0$ above $\rho_{\kappa_0})$ if $\kappa_0 + 1 = p$, and $\boxed{\#\mid \rho_{\kappa_0}\mid \#}$ $(r'_{\kappa_0+1}$ above $\rho_{\kappa_0})$ otherwise;

- if $\rho^+ = \{l'\}$: $\boxed{\#\mid \#\mid \rho_{\kappa_0}}$ $(l'_0$ above $\rho_{\kappa_0})$ if $\kappa_0 + 1 = q$, and $\boxed{\#\mid \rho_{\kappa_0}\mid \#}$ $(l'_{\kappa_0+1}$ above $\rho_{\kappa_0})$ otherwise;

- if $\rho^+ = \{z'\}$: $\boxed{\#\mid \rho_{\kappa_0}\mid \#}$ $(z'_\theta$ above $\rho_{\kappa_0})$.

If the collision emits more than one signal, it progresses until a signal leaves:

$\forall k,\ \kappa_0 \le k < \kappa_1$: $\boxed{\#\mid \rho_k\mid \#}$ $(\rho_{k+1}$ above $\rho_k)$.

If the collision emits two signals as illustrated in Fig. 9(b):

- if $\rho^+ = \{z', r'\}$, add: $\boxed{\#\mid \rho_{\kappa_1}\mid \#}$ $(z'_\theta$ above $\rho_{\kappa_1})$ and $\boxed{\rho_{\kappa_1}\mid \#\mid \#}$ $(r'_0$ above $\rho_{\kappa_1})$;

- if $\rho^+ = \{l', z'\}$, add: $\boxed{\#\mid \#\mid \rho_{\kappa_1}}$ $(l'_0$ above $\rho_{\kappa_1})$ and $\boxed{\#\mid \rho_{\kappa_1}\mid \#}$ $(z'_\theta$ above $\rho_{\kappa_1})$;

- if $\rho^+ = \{l', r'\}$,

  - if $q < p$ ($q < p$ is symmetrical), add: $\boxed{\#\mid \#\mid \rho_{\kappa_1}}$ $(l'_0$ above $\rho_{\kappa_1})$ and $\boxed{\#\mid \rho_{\kappa_1}\mid \#}$ $(r'_{\kappa_1+1}$ above $\rho_{\kappa_1})$.

– if $q = p = 1$, add: 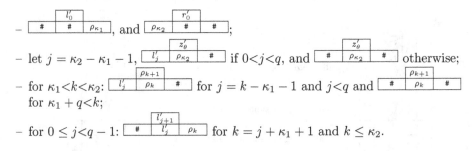 and .

If the collision emits three signals, in the case $q \leq p$ (the rest is symmetric) as illustrated in Fig. 9(c), the following is added:

– , and ;

– let $j = \kappa_2 - \kappa_1 - 1$, if $0 < j < q$, and otherwise;

– for $\kappa_1 < k < \kappa_2$: for $j = k - \kappa_1 - 1$ and $j < q$ and for $\kappa_1 + q < k$;

– for $0 \leq j < q - 1$: for $k = j + \kappa_1 + 1$ and $k \leq \kappa_2$.

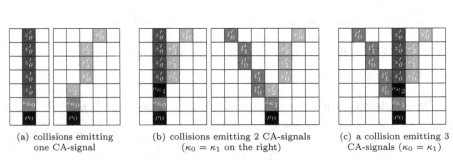

(a) collisions emitting one CA-signal

(b) collisions emitting 2 CA-signals ($\kappa_0 = \kappa_1$ on the right)

(c) a collision emitting 3 CA-signals ($\kappa_0 = \kappa_1$)

Fig. 9. Collisions and CA-Signals getting away from each other with $p = 3$ and $q = 2$.

After the collision, signals move away from each other as illustrated in Fig. 9. This is handled with the transitions:

– $\forall k,\ 0 \leq k < p - 1$: and ,

– $\forall k,\ 0 \leq k < q - 1$: and , and

– $\forall k, m,\ 0 \leq m < q - 1$ and $0 \leq k < p - 1$: .

A scale is used to ensure that the continuous mesh is correctly discretized: spaces between stationary signals should be integers as well as times of collisions. From the definition of encounter, any (positive) multiple of $p + q$ is enough. Let $\delta = 3(p + q)$ be the *discretization scale*.

Let c be an initial configuration of the discretized SM. The corresponding *initial CA-configuration*, c_0, is defined by: $\forall i \in \mathbb{Z}$, $c_0(i) = \psi(c(i/\delta))$ where $\psi(\oslash) = \#$, $\psi(z) = z_\theta$ for any stationary meta-signal, and $\psi(\mu) = \mu_0$ for any other. This ensures the validity of the phase hypothesis at the first round of collisions (Fig. 10).

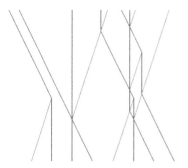 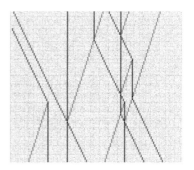

Fig. 10. A continuous space-time diagram and the generated CA discretization.

5   Correctness

In this section, $q \leq p$ is assumed (the rest follows by symmetry).

A discrete mesh is the discrete counterpart of the continuous meshes as illustrated in Fig. 11. The (δ, p, q, n)-CA-Mesh corresponds to the union of:

- $V_v = \{3v\} \times \mathbb{N}$ with $v \in \{0, 1, 2, \ldots, n(q + p)\}$, $(3 = \delta/(q + p))$,
- $L_l = \left\{ (x, t) \,\middle|\, 0 \leq t \vee x \leq n\delta \vee x = \left\lceil \frac{l\delta - t}{q} \right\rceil \right\}$ with $l \in \mathbb{N}$,
- $R_r = \left\{ (x, t) \,\middle|\, 0 \leq t \vee 0 \leq x \vee x = \left\lfloor \frac{t - r\delta}{p} \right\rfloor \right\}$ with $r \in \{(-n.p), \ldots, -1, 0, 1, \ldots\}$.

Each set corresponds to a CA-signal (not necessary issued from the initial configuration). They can overlap for consecutive iterations (as in Figs. 9 and 12).

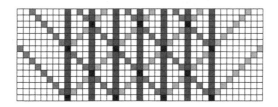

Fig. 11. The (4,1,1,3)-CA-Mesh ($\delta = 8 = 4(1 + 1)$).

In the $(\delta = 3(p+q), p, q, n)$-(CA-)mesh, the *CA-encounter* of coordinates (v, r), c_v^r, corresponds to: $\left(\left\lfloor \frac{\delta}{p+q} v + \frac{q}{p+q} \right\rfloor, \left\lfloor r\delta + \frac{p\delta}{p+q} v + \frac{pq}{p+q} \right\rfloor \right) = (3v, 3pv + r\delta + \kappa_0)$ where the (v, r) coordinates are integers such that $0 \leq v \leq n(p+q)$ and $-n.q \leq r$. The last term in the formula comes from the phases.

Figure 12 depicts an entire (CA-)collision. Under the hypothesis that the κ_0 site is located at some e_v^r, ρ_0 has coordinates $(3v, 3pv + r\delta)$ and thus corresponds to the SM-encounter e_v^r at scale δ. This relates discrete and continuous meshes.

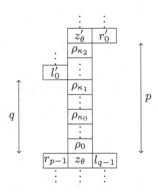

Fig. 12. Entire generic collision $\rho = \{r, z, l\} \rightarrow \{l', z', r'\}$.

With the same hypothesis, the following holds: ρ_0 belongs to V_v, R_r and L_{v+r}, ρ_{κ_2} belongs to R_r, and ρ_{κ_1} belongs to L_{v+r}. This fixes all three CA-signals over a period. The whole collision is in the (δ, p, q, n)-CA-Mesh. As can be seen in Fig. 12, meta-signals extend out of the collision from both ends so that CA-signals directly connect into.

The hypothesis is always satisfied as can be proved by induction: CA-signals in the initial configuration are in the CA-Mesh with the right phases and if CA-signals from the mesh collide, so are the collisions and the resulting CA-signals. Altogether, it proves that:

Theorem 1. *All the non-quiescent sites of the run from the simulating CA-configuration are located inside the (δ, p, q, n)-CA-Mesh. Up to scaling, this mesh coincides with the continuous (p, q, n)-SM-mesh.*

The $\mathfrak{A}$-$\mathcal{A}$ representation relation is defined straightforwardly from the construction. Let l, z and r be any leftward, stationary and rightward meta-signals:

$$\forall i, 0 \leq i < q, l \mathcal{R} l_i \qquad z \mathcal{R} z_\theta \qquad \forall i, 0 \leq i < p, r \mathcal{R} r_i \ .$$

Let $\rho = \{r, z, l\} \rightarrow \{l', z', r'\}$ be any rule.

$$\forall i, 0 \leq i < \kappa_0, r \mathcal{R} \rho_i, \qquad z \mathcal{R} \rho_i, \text{ and } l \mathcal{R} \rho_i,$$
$$\rho \mathcal{R} \rho_{\kappa_0},$$
$$\forall i, \kappa_0 \leq i < \kappa_1, r' \mathcal{R} \rho_i, \qquad z' \mathcal{R} \rho_i, \text{ and } l' \mathcal{R} \rho_i,$$
$$\forall i, \kappa_1 < i \leq \kappa_2, r' \mathcal{R} \rho_i \text{ and } z' \mathcal{R} \rho_i.$$

A structural induction on the mesh proves that the SM and CA LDAG's are identical, so that:

Theorem 2. *Any 3-speed rational signal machine can be simulated exactly by a cellular automaton on any infinite run preserving the geometry.*

6 Conclusion

The exact conversion of a 3-speed rational signal machine into a cellular automata has been proved and implemented in Java. The construction relies on preserving a dynamics that is trapped inside a discrete-like mesh. Moreover this is valid on any infinite run and preserves the relative position of collisions.

This result is tight: as soon as either irrationality [2] or four speeds is allowed, Zeno-like phenomena (infinitely many collisions in finite duration) is possible and there is no hope for exact discretization. Even if the techniques presented here can be extended to any number of (rational) speeds, there is no information on how long the discretization remains valid. We believe that testing the validity for a given amount of time is as complex as running the signal machine and its simulation.

The construction presented here naturally extends to any SM computation that is constrained to remain on a mesh (we do not expect such a property to be decidable whether the mesh is provided or not in the general case). One perspective is to either work with some extra piece of information proving the containment into a mesh or the non-existence of an accumulation (which is highly non decidable [8]) or other problematic behavior. One important step would be to discretize rational signal machines with four or more speeds known to be Turing complete without accumulating as in [11].

Another perspective is to provide approximation. The quality of the approximation would have to be defined and if possible a bound guaranteed.

References

1. Adamatzky, A. (ed.): Collision Based Computing. Springer, London (2002)
2. Becker, F., Chapelle, M., Durand-Lose, J., Levorato, V., Senot, M.: Abstract geometrical computation 8: Small machines, accumulations & rationality (2013, submitted). http://arxiv.org/abs/1307.6468
3. Blum, L., Shub, M., Smale, S.: On a theory of computation and complexity over the real numbers: NP-completeness, recursive functions and universal machines. Bull. Amer. Math. Soc. **21**(1), 1–46 (1989)
4. Cook, M.: Universality in elementary cellular automata. Complex Syst. **15**, 1–40 (2004)
5. Crutchfield, J.P., Mitchell, M., Das, R.: The evolutionary design of collective computation in cellular automata. In: Crutchfield, J.P., Schuster, P.K. (eds.) Evolutionary Dynamics-Exploring the Interplay of Selection, Neutrality, Accident, and Function, pp. 361–411. Oxford University Press, New York (2003)
6. Durand-Lose, J.: Reversible space-time simulation of cellular automata. Theor. Comp. Sci. **246**(1–2), 117–129 (2000)
7. Durand-Lose, J.: Calculer géométriquement sur le plan - machines à signaux. Habilitation à Diriger des Recherches, École Doctorale STIC, Université de Nice-Sophia Antipolis (2003, in French)
8. Durand-Lose, J.: Forecasting black holes in abstract geometrical computation is highly unpredictable. In: Cai, J.-Y., Cooper, S.B., Li, A. (eds.) TAMC 2006. LNCS, vol. 3959, pp. 644–653. Springer, Heidelberg (2006)

9. Durand-Lose, J.: Reversible conservative rational abstract geometrical computation is Turing-universal. In: Beckmann, A., Berger, U., Löwe, B., Tucker, J.V. (eds.) CiE 2006. LNCS, vol. 3988, pp. 163–172. Springer, Heidelberg (2006)

10. Durand-Lose, J.: Abstract geometrical computation and the linear Blum, Shub and Smale Model. In: Cooper, S.B., Löwe, B., Sorbi, A. (eds.) CiE 2007. LNCS, vol. 4497, pp. 238–247. Springer, Heidelberg (2007)

11. Durand-Lose, J.: Abstract geometrical computation 4: small Turing universal signal machines. Theor. Comp. Sci. **412**, 57–67 (2011)

12. Durand-Lose, J.: Irrationality is needed to compute with signal machines with only three speeds. In: Bonizzoni, P., Brattka, V., Löwe, B. (eds.) CiE 2013. LNCS, vol. 7921, pp. 108–119. Springer, Heidelberg (2013)

13. Fischer, P.C.: Generation of primes by a one-dimensional real-time iterative array. J. ACM **12**(3), 388–394 (1965)

14. Goto, E.: Ōtomaton ni kansuru pazuru [Puzzles on automata]. In: Kitagawa, T. (ed.) Jōhōkagaku eno michi [The Road to information science], pp. 67–92. Kyoristu Shuppan Publishing Co., Tokyo (1966)

15. Mazoyer, J., Rapaport, I.: Inducing an order on cellular automata by a grouping operation. In: Meinel, C., Morvan, M. (eds.) STACS 1998. LNCS, vol. 1373, pp. 116–127. Springer, Heidelberg (1998)

16. Mazoyer, J., Terrier, V.: Signals in one-dimensional cellular automata. Theor. Comp. Sci. **217**(1), 53–80 (1999)

17. Neumann, J.: Theory of Self-Reproducing Automata. University of Illinois Press, Urbana (1966)

18. Ollinger, N., Richard, G.: Four states are enough!. Theor. Comp. Sci. **412**(1–2), 22–32 (2011)

19. Richard, G.: Systèmes de particules et collisions discrètes dans les automates cellulaires. Ph.D. thesis, Aix-Marseille Université (2008)

20. Yunès, J.-B.: Simple new algorithms which solve the firing squad synchronization problem: a 7-states 4n-steps solution. In: Durand-Lose, J., Margenstern, M. (eds.) MCU 2007. LNCS, vol. 4664, pp. 316–324. Springer, Heidelberg (2007)

An "almost dual" to Gottschalk's Conjecture

Silvio Capobianco[1(✉)], Jarkko Kari[2], and Siamak Taati[3]

[1] Institute of Cybernetics at Tallinn University of Technology, Tallinn, Estonia
silvio@cs.ioc.ee
[2] Department of Mathematics and Statistics, University of Turku, Turku, Finland
jkari@utu.fi
[3] Mathematical Institute, Leiden University, Leiden, The Netherlands
siamak.taati@gmail.com

Abstract. We discuss cellular automata over arbitrary finitely generated groups. We call a cellular automaton post-surjective if for any pair of asymptotic configurations, every pre-image of one is asymptotic to a pre-image of the other. The well known dual concept is pre-injectivity: a cellular automaton is pre-injective if distinct asymptotic configurations have distinct images. We prove that pre-injective, post-surjective cellular automata are reversible. We then show that on sofic groups, where it is known that injective cellular automata are surjective, post-surjectivity implies pre-injectivity. As no non-sofic groups are currently known, we conjecture that this implication always holds. This mirrors Gottschalk's conjecture that every injective cellular automaton is surjective.

Keywords: Cellular automata · Reversibility · Sofic groups

1 Introduction

Cellular automata (briefly, CA) are parallel synchronous systems on regular grids where the next state of a point depends on the current state of a finite neighborhood. The grid is determined by a finitely generated group and can be visualized as the Cayley graph of the group. In addition to being a useful tool for simulations, CA are studied as models of massively parallel computers, and as dynamical systems on symbolic spaces. From a combinatorial point of view, interesting questions arise as to how the properties of the global transition function (obtained by synchronous application of the local update rule at each point) are related to one another.

One such relation is provided by Bartholdi's theorem [1], stating that *amenable* groups (*i.e.*, those which have a finitely additive probability measure,

S. Capobianco—This research was supported by the ERDF funded project Coinduction, the Estonian Ministry of Education and Research institutional research grant IUT33-13, and the Estonian Science Foundation grant no. 9398.
S. Taati—The work of ST is supported by ERC Advanced Grant 267356-VARIS of Frank den Hollander.

© IFIP International Federation for Information Processing 2016
Published by Springer International Publishing Switzerland 2016. All Rights Reserved
M. Cook and T. Neary (Eds.): AUTOMATA 2016, LNCS 9664, pp. 77–89, 2016.
DOI: 10.1007/978-3-319-39300-1_7

defined on every subset, and invariant by multiplication on the left) are precisely those where the *Garden of Eden theorem* holds. The latter states that surjective CA are *pre-injective*, that is, two configurations differing only in finitely many points have equal image only if they are equal. By [7, Theorem 4.7], the Garden of Eden theorem still holds for CA on subshifts that are of finite type and are strongly irreducible. Counterexamples with generic subshifts are known already in dimension 1. Furthermore, bijectivity is always equivalent to *reversibility*, that is, the existence of an inverse that is itself a CA.

A very remarkable consequence of the Garden of Eden theorem is that amenable groups are *surjunctive*: that is, every CA on an amenable group, which is injective on the *full shift* of all the possible configurations, is surjective. On the other hand, it is easy to prove that the free group on two generators, which is the main example of non-amenable group, is also surjunctive. Indeed, at the present time, not a single example of injective, non-surjective CA is known! This led Gottschalk to conjecture, in his 1973 paper [8], that *all* groups are actually surjunctive. The conjecture is known to hold for the class of *sofic groups*, originally defined by Gromov in the context of geometric group theory. Remarkably, no examples of non-sofic groups are known at the present time.

In this paper, which expands our previous work from Automata 2015 [4], we discuss *post-surjectivity*, a parallel property to pre-injectivity, which we define as follows: however given a configuration c and a preimage e, every configuration c' asymptotic to c has a pre-image e' asymptotic to e. While pre-injectivity is *weaker* than injectivity, post-surjectivity turns out to be *stronger* than surjectivity. It is natural to ask whether such trade-off between injectivity and surjectivity preserves bijectivity.

First, we prove that post-surjectivity and pre-injectivity together imply reversibility: that is, the trade-off above actually holds over all groups. Next, we show that, in the context of sofic groups, post-surjectivity actually implies pre-injectivity. From all this we formulate an "almost dual" to Gottschalk's conjecture, that every post-surjective CA is pre-injective—or, equivalently, reversible.

2   Background

Let X be a set. We indicate by $\mathcal{PF}(X)$ the collection of all finite subsets of X. If X is finite, we indicate by $|X|$ the number of its elements.

Let $\mathbb{G}$ be a group and let $U, V \subseteq \mathbb{G}$. We put $UV = \{x \cdot y \mid x \in U, y \in V\}$, and $U^{-1} = \{x^{-1} \mid x \in U\}$. If $U = \{g\}$ we write gV for $\{g\}V$.

A *labeled graph* is a triple (V, L, E) where V is a set of *vertices*, L is a set of *labels*, and $E \subseteq V \times L \times V$ is a set of *labeled edges*. A *labeled graph isomorphism* from (V_1, L, E_1) to (V_2, L, E_2) is a bijection $\phi : V_1 \to V_2$ such that, for every $x, y \in V_1$ and $\ell \in L$, $(x, \ell, y) \in E_1$ if and only if $(\phi(x), \ell, \phi(y)) \in E_2$. We will sometimes say that (V, E) is an L-labeled graph to mean that (V, L, E) is a labeled graph.

A subset B of $\mathbb{G}$ is a *set of generators* for $\mathbb{G}$ if every $g \in \mathbb{G}$ can be written as $g = x_0 \cdots x_{n-1}$ for suitable $n \geq 0$ and $x_0, \ldots, x_{n-1} \in B \cup B^{-1}$. The group $\mathbb{G}$ is *finitely generated* (briefly, f.g.) if B can be chosen finite.

Let B be a finite set of generators for the group $\mathbb{G}$. The *Cayley graph* of $\mathbb{G}$ w.r.t. B is the $(B \cup B^{-1})$-labeled graph $(\mathbb{G}, E)$ where $E = \{(g, x, h) \mid gx = h\}$. The *length* of $g \in \mathbb{G}$ with respect to B is the *minimum* length $n = \|g\|_B$ of a representation $g = x_0 \cdots x_{n-1}$. The *distance* between g and h with respect to B is $d_B(g, h) = \|g^{-1} \cdot h\|_B$, i.e., the length of the shortest path from g to h in the Cayley graph of $\mathbb{G}$ with respect to B. The *disk* of center g and radius r with respect to B is the set $D_{B,r}(g)$ of those $h \in \mathbb{G}$ such that $d_B(g, h) \leq r$. We omit g if it is the identity element $1_{\mathbb{G}}$ of $\mathbb{G}$ and write $D_{B,r}$ for $D_{B,r}(1_{\mathbb{G}})$. The distance between two subsets $U, V \subseteq \mathbb{G}$ is $d_B(U, V) = \inf\{d_B(u, v) \mid u \in U, v \in V\}$. We omit B if irrelevant or clear from the context.

A group $\mathbb{G}$ is *amenable* if for every $K \in \mathcal{PF}(\mathbb{G})$ and every $\varepsilon > 0$ there exists a nonempty $F \in \mathcal{PF}(\mathbb{G})$ such that $|F \cap kF| > (1 - \varepsilon)|F|$ for every $k \in K$. The groups $\mathbb{Z}^d$ are amenable, whereas the *free groups* on two or more generators are not. For an introduction to amenability see, *e.g.*, [5, Chapter 4].

Let S be a finite set and let $\mathbb{G}$ be a group. The elements of the set $S^{\mathbb{G}}$ are called *configurations*. The space $S^{\mathbb{G}}$ is given the *product topology* by considering S as a discrete set. This makes $S^{\mathbb{G}}$ a compact space by Tychonoff's theorem. In the prodiscrete topology, two configurations are "near" if they coincide on a "large" finite subset of $\mathbb{G}$. Indeed, if B is a finite set of generators for $\mathbb{G}$, then setting $d_B(c, e) = 2^{-n}$, where n is the smallest $r \geq 0$ such that c and e differ on $D_{B,r}$, defines a distance that induces the prodiscrete topology. Given $c, c' \in S^{\mathbb{G}}$, we call $\Delta(c, c') = \{g \in \mathbb{G} \mid c(g) \neq c'(g)\}$ the *difference set* of c and c'. Two configurations are *asymptotic* if they differ at most on finitely many points of $\mathbb{G}$. A *pattern* is a function $p : E \to S$ where E is a finite subset of $\mathbb{G}$.

For $g \in \mathbb{G}$, the *translation* by g is the function $\sigma_g : S^{\mathbb{G}} \to S^{\mathbb{G}}$ that sends an arbitrary configuration c into the configuration $\sigma_g(c)$ defined by

$$\sigma_g(c)(x) = c(g \cdot x) \quad \forall x \in \mathbb{G}. \tag{1}$$

A *shift subspace* (briefly, *subshift*) is a subset X of $S^{\mathbb{G}}$ which is closed (equivalently, compact) and invariant by translation. The set $S^{\mathbb{G}}$ itself is referred to as the *full shift*. It is well known (cf. [12]) that every subshift is determined by a set of *forbidden patterns*, in the sense that the elements of the subshift are precisely those configurations in which the translations of the forbidden patterns do not occur. If such set can be chosen finite, X is called a *shift of finite type* (briefly, SFT). A pattern that appears on some configuration in X is said to be *admissible* for X. The set of patterns that are admissible for X is called the *language* of X, indicated as $\mathcal{L}_X$.

A *cellular automaton* (briefly, CA) on a group $\mathbb{G}$ is a triple $\mathcal{A} = \langle S, \mathcal{N}, f \rangle$ where the *set of states* S is finite and has at least two elements, the *neighborhood* $\mathcal{N}$ is a finite subset of $\mathbb{G}$, and the *local update rule* is a function that associates to every pattern $p : \mathcal{N} \to S$ a state $f(p) \in S$. The *global transition function* of $\mathcal{A}$ is the function $F_{\mathcal{A}} : S^{\mathbb{G}} \to S^{\mathbb{G}}$ defined by

$$F_{\mathcal{A}}(c)(g) = f\left((\sigma_g(c))|_{\mathcal{N}}\right) \quad \forall g \in \mathbb{G}, \tag{2}$$

that is, if $\mathcal{N} = \{n_1, \ldots, n_m\}$, then $F_{\mathcal{A}}(c)(g) = f(c(g \cdot n_1), \ldots, c(g \cdot n_m))$. Observe that (2) is continuous in the prodiscrete topology and commutes with the translations, *i.e.*, $F_\mathcal{A} \circ \sigma_g = \sigma_g \circ F_\mathcal{A}$ for every $g \in \mathbb{G}$. The *Curtis-Hedlund-Lyndon theorem* states that the continuous and translation-commuting functions from $S^\mathbb{G}$ to itself are precisely the CA global transition functions.

We shall use the following notation to represent the application of the local rule on patterns. If $p : E \to S$ and $q : C \to S$ are two patterns, we write $p \xrightarrow{f} q$ to indicate that $C\mathcal{N} \subseteq E$ and $q(g) = f\left((\sigma_g(p))|_\mathcal{N}\right)$ for each $g \in C$.

If X is a subshift and $F_\mathcal{A}$ is a cellular automaton, it is easy to see that $F_\mathcal{A}(X)$ is also a subshift. If, in addition, $F_\mathcal{A}(X) \subseteq X$, we say that $\mathcal{A}$ is a CA on the subshift X. From now on, when we speak of cellular automata on $\mathbb{G}$ without specifying any subshift, we will imply that such subshift is the full shift.

We may refer to injectivity, surjectivity, etc. of the cellular automaton $\mathcal{A}$ on the subshift X meaning the corresponding properties of $F_\mathcal{A}$ when restricted to X. From basic facts about compact spaces, it follows that the inverse of the global transition function of a bijective cellular automaton $\mathcal{A}$ is itself the global transition function of some cellular automaton. In this case, we say that $\mathcal{A}$ is *reversible*. A group $\mathbb{G}$ is *surjunctive* if for every finite set S, every injective cellular automaton on the full shift $S^\mathbb{G}$ is surjective. Currently, there are no known examples of non-surjunctive groups.

Conjecture 1 (Gottschalk [8]). Every group is surjunctive.

If $\mathbb{G}$ is a subgroup of Γ and $\mathcal{A} = \langle S, \mathcal{N}, f \rangle$ is a cellular automaton on $\mathbb{G}$, the cellular automaton $\mathcal{A}^\Gamma$ *induced* by $\mathcal{A}$ on Γ has the same set of states, neighborhood, and local update rule as $\mathcal{A}$, and maps S^Γ (instead of $S^\mathbb{G}$) into itself via $F_{\mathcal{A}^\Gamma}(c)(\gamma) = f\left(c(\gamma \cdot n_1), \ldots, c(\gamma \cdot n_m)\right)$ for every $\gamma \in \Gamma$. We also say that $\mathcal{A}$ is the *restriction* of $\mathcal{A}^\Gamma$ to $\mathbb{G}$. In addition, if $X \subseteq S^\mathbb{G}$ is a subshift defined by a set F of forbidden patterns on $\mathbb{G}$, then the subshift $X^\Gamma \subseteq S^\Gamma$ obtained from the same set F of forbidden patterns satisfies the following property: if $\mathcal{A}$ is a CA on X, then $\mathcal{A}^\Gamma$ is a CA on X^Γ, and vice versa. (Here, it is fundamental that all the forbidden patterns have their supports in $\mathbb{G}$.) It turns out (cf. [5, Section 1.7] or [2, Theorem 5.3]) that injectivity and surjectivity are preserved by both induction and restriction.

Let $\mathcal{A} = \langle S, \mathcal{N}, f \rangle$ be a CA on a subshift X, let $p : E \to S$ be an admissible pattern for X, and let $E\mathcal{N} \subseteq M \in \mathcal{PF}(\mathbb{G})$. A *pre-image* of p on M under $\mathcal{A}$ is a pattern $q : M \to S$ that is admissible for X such that $q \xrightarrow{f} p$. An *orphan* is an admissible pattern that has no admissible pre-image, or equivalently, a pattern that is admissible for X but not admissible for $F_\mathcal{A}(X)$. Similarly, a configuration which is not in the image of X by $F_\mathcal{A}$ is a *Garden of Eden* for $\mathcal{A}$. By a compactness argument, every Garden of Eden contains an orphan. We call this the *orphan pattern principle*. A cellular automaton $\mathcal{A}$ is *pre-injective* if every two asymptotic configurations c, e satisfying $F_\mathcal{A}(c) = F_\mathcal{A}(e)$ are equal. The *Garden of Eden theorem* (cf. [6]) states that, for CA on amenable groups, pre-injectivity is equivalent to surjectivity; on non-amenable groups, the two properties appear to be independent of each other.

Definition 1. *Let $\mathbb{G}$ be a finitely generated group, let B be a finite set of generators for $\mathbb{G}$, and let S be a finite set. A subshift $X \subseteq S^\mathbb{G}$ is strongly* irreducible *if there exists* $r \geq 0$ *such that, for every two admissible patterns* $p_1 : E_1 \to S, p_2 : E_2 \to S$ *such that* $d_B(E_1, E_2) \geq r$, *there exists* $c \in X$ *such that* $c|_{E_1} = p_1$ *and* $c|_{E_2} = p_2$. *We then say that* r *is a* constant of strong irreducibility *for* X *with respect to* B.

The notion of strong irreducibility does not depend on the choice of the finite set of generators, albeit the associated constant of strong irreducibility usually does. If no ambiguity is possible, we will suppose B fixed once and for all, and always speak of r relative to B. For $\mathbb{G} = \mathbb{Z}$, strong irreducibility is equivalent to existence of $r \geq 0$ such that, for every two $u, v \in \mathcal{L}_X$, there exists $w \in S^r$ satisfying $uwv \in \mathcal{L}_X$.

As a consequence of the definition, strongly irreducible subshifts are *mixing*: given two open sets $U, V \subseteq X$, the set of those $g \in \mathbb{G}$ such that $U \cap \sigma_g(V) = \emptyset$ is, at most, finite. In addition to this, as by [12, Theorem 8.1.16], the Garden of Eden theorem is still valid on strongly irreducible subshifts. We remark that for one-dimensional subshifts of finite type, strong irreducibility is equivalent to the mixing property.

Another property of strongly irreducible subshifts, which will have a crucial role in the next section, is that they allow a "cut and paste" technique which is very common in proofs involving the full shift, but may be inapplicable for more general shifts.

Proposition 1. *Let* $X \subseteq S^{\mathbb{G}}$ *be a strongly irreducible subshift, let* $c \in X$, *and let* $p : E \to S$ *be an admissible pattern for* X. *There exists* $c' \in X$ *asymptotic to* c *such that* $c'|_E = p$.

Proof. It is not restrictive to suppose $E = D_n$ for suitable $n \geq 0$. Let $r \geq 0$ be a constant of strong irreducibility for X. Writing $E_k = D_{n+r+k} \setminus D_{n+r}$ for $k \geq 1$, we have of course $d(E, E_k) = r$. Set $p_k = c|_{E_k}$. By strong irreducibility, there exists $c_k \in X$ such that $c_k|_E = p$ and $c_k|_{E_k} = p_k$. Then every limit point c' of $\{c_k\}_{k \geq 1}$, which exists and belongs to X because of compactness, satisfies the thesis. $\square$

3   Post-surjectivity

The notion of post-surjectivity is a sort of "dual" to pre-injectivity: it is a strengthening of surjectivity, in a similar way that pre-injectivity is a weakening of injectivity. The maps that are both pre-injective and post-surjective were studied in [11] under the name of complete pre-injective maps.

Definition 2. *Let* $\mathbb{G}$ *be a group,* S *a finite set, and* $X \subseteq S^{\mathbb{G}}$ *a strongly irreducible shift of finite type. A cellular automaton* $\mathcal{A} = \langle S, \mathcal{N}, f \rangle$ *on* X *is* post-surjective *if, however given* $c \in X$ *and a predecessor* $e \in X$ *of* c, *every configuration* $c' \in X$ *asymptotic to* c *has a predecessor* $e' \in X$ *asymptotic to* e.

When $X = S^{\mathbb{G}}$ is the full shift, if no ambiguity is present, we will simply say that the CA is post-surjective.

Example 1. Every reversible cellular automaton is post-surjective. If $R \geq 0$ is a neighborhood radius for the inverse CA, and c and c' coincide outside D_N, then their unique pre-images e and e' must coincide outside D_{N+R}.

Example 2. The xor CA with the right-hand neighbor (the one-dimensional elementary CA with rule 102) is surjective, but not post-surjective. As the xor function is a permutation of each of its arguments given the other, every $c \in \{0,1\}^{\mathbb{Z}}$ has two pre-images, uniquely determined by their value in a single point. However (actually, because of this!) $\ldots 000 \ldots$ is a fixed point, but $\ldots 010 \ldots$ only has pre-images that take value 1 infinitely often.

The qualification "post-surjective" is well earned:

Proposition 2. *Let $X \subseteq S^{\mathbb{G}}$ be a strongly irreducible* SFT. *Every post-surjective* CA *on X is surjective.*

Proof. Let $r \geq 0$ be the constant of strong irreducibility of X, i.e., let every two admissible patterns whose supports have distance at least r be jointly sub-patterns of some configuration. Take an arbitrary $e \in X$ and set $c = F(e)$. Let $p : E \to S$ be an admissible pattern for X. By Proposition 1, there exists $c' \in X$ asymptotic to c such that $c'|_E = p$. By post-surjectivity, such c' has a pre-image in X, which means p has a pre-image admissible for X. The thesis follows from the orphan pattern principle.   □

From Proposition 2 together with [7, Theorem 4.7] follows:

Proposition 3. *Let $\mathbb{G}$ be an amenable group and let $X \subseteq S^{\mathbb{G}}$ be a strongly irreducible* SFT. *Every post-surjective* CA *on X is pre-injective.*

In addition, via a reasoning similar to the one employed in [5, Sect. 1.7] and [3, Remark 18], we can prove:

Proposition 4. *Let $\mathcal{A} = \langle S, \mathcal{N}, f \rangle$ be a cellular automaton on the group $\mathbb{G}$, let Γ be a group that contains $\mathbb{G}$, and let $\mathcal{A}^{\Gamma}$ be the* CA *induced by $\mathcal{A}$ on Γ. Then $\mathcal{A}$ is post-surjective on $S^{\mathbb{G}}$ if and only if $\mathcal{A}^{\Gamma}$ is post-surjective on S^{Γ}.*

In particular, post-surjectivity of arbitrary CA is equivalent to post-surjectivity on the subgroup generated by the neighborhood.

Proof. Suppose that $\mathcal{A}$ is post-surjective. Let J be a set of representatives of the left cosets of $\mathbb{G}$ in Γ, i.e., let $\Gamma = \bigsqcup_{j \in J} j\mathbb{G}$. Let $c, c' \in S^{\Gamma}$ two asymptotic configurations and let e be a pre-image of c. For every $j \in J$ and $g \in \mathbb{G}$ set

$$c_j(g) = c(j \cdot g);$$
$$c'_j(g) = c'(j \cdot g);$$
$$e_j(g) = e(j \cdot g).$$

By construction, c_j is asymptotic to c'_j and has e_j as a pre-image according to $\mathcal{A}$. Moreover, as c and c' are asymptotic in the first place, $c'_j \neq c_j$ only for finitely many $j \in J$. For every $j \in J$ let $e'_j \in S^G$ be a pre-image of c'_j according to $\mathcal{A}$ asymptotic to e_j, if $c'_j \neq c_j$, and e_j itself if $c'_j = c_j$. Then,

$$e'(\gamma) = e_j(g) \quad \Longleftrightarrow \quad \gamma = j \cdot g$$

defines a pre-image of c' asymptotic to e according to $\mathcal{A}^\Gamma$.

The converse implication is immediate. ☐

Example 3. Let $X \subseteq S^{\mathbb{Z}}$ be a one-dimensional strongly irreducible SFT and let $\mathcal{A} = \langle S, \mathcal{N}, f \rangle$ be a post-surjective CA on X. Then $\mathcal{A}$ is reversible.

Suppose it is not so. For CA on one-dimensional strongly irreducible SFT, reversibility is equivalent to injectivity on periodic configurations. Namely, if two distinct configurations with the same image exist, then one can construct two distinct *periodic* configurations with the same image. Let then $u, v, w \in S^*$ be such that $e_u = \ldots uuu \ldots$, the configuration obtained by extending u periodically in both directions, and $e_v = \ldots vvv \ldots$ are different and have the same image $c = \ldots www \ldots$. It is not restrictive to suppose $|u| = |v| = |w|$. Without loss of generality, we also assume that X is defined by a set of forbidden words of length at most $|u|$.

Let $r \geq 0$ be a strong irreducibility constant for X and let $p, q \in S^r$ be such that $upv, vqu \in \mathcal{L}_X$. The two configurations $c_{u,v} = F(\ldots uupvv \ldots)$ and $c_{v,u} = F(\ldots vvquu \ldots)$ are both asymptotic to c. By post-surjectivity, there exist $x, y \in \mathcal{L}_X$ such that $e_{u,v} = \ldots uuxvv \ldots$ and $e_{v,u} = \ldots vvyuu \ldots$ satisfy $F(e_{u,v}) = F(e_{v,u}) = c$. Again, it is not restrictive to suppose that $|x| = |y| = m \cdot |u|$ for some $m \geq 1$, and that x and y start in $e_{u,v}$ and $e_{v,u}$ at the same point $i \in \mathbb{Z}$.

Let us now consider the configuration $e' = \ldots uuxv^N yuu \ldots$. By our previous discussion, for N large enough (*e.g.*, so that x and y do not have overlapping neighborhoods) $F_{\mathcal{A}}(e')$ cannot help but be c. Now, recall that e_u is also a pre-image of c and note that e_u and e' are asymptotic but distinct. Then $\mathcal{A}$ is surjective, but not pre-injective, contradicting the Garden of Eden theorem [12, Theorem 8.1.16] (Fig. 1).

Example 3 depends critically on dimension 1, where CA that are injective on periodic configurations are reversible. Moreover, in our final step, we invoke the Garden of Eden theorem, which we know from [1,6] not to hold for CA on generic groups. Not all is lost, however: maybe, by explicitly adding the pre-injectivity requirement, we can recover Example 3 on more general groups?

It turns out that it is so, at least for CA on full shifts. To see this, we need a preliminary lemma.

Lemma 1. *Let $\mathcal{A}$ be a post-surjective CA on a finitely generated group $\mathbb{G}$ and let F be its global transition function. There exists $N \geq 0$ such that, given any three configurations c, c', e with $c = F(e)$ and $\Delta(c, c') = \{1_{\mathbb{G}}\}$, there exists a pre-image e' of c' which coincides with e outside D_N.*

Proof. By contradiction, assume that for every $n \geq 0$ there exist $c_n \in S^{\mathbb{G}}, e_n \in F^{-1}(c_n)$, and $c'_n \neq c_n$ such that $\Delta(c_n, c'_n) = \{1_{\mathbb{G}}\}$, but every $e'_n \in F^{-1}(c'_n)$ differs

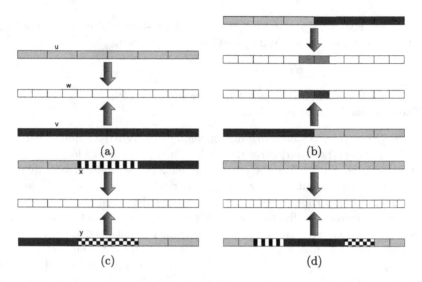

Fig. 1. A graphical description of the argument in Example 3 for the full shift. (a) Let a 1D periodic configuration w have two different (periodic) preimages u and v. (b) By swapping the right-hand halves of the preimages, the new images only differ from the initial one in finitely many points. (c) By post-surjectivity, we can change them in finitely many points, and get two preimages of the initial configuration. (d) Then a violation of the Garden of Eden theorem occurs.

from e_n on some point outside D_n. By compactness, there exits a sequence n_i such that the limits $c = \lim_{i\to\infty} c_{n_i}, c' = \lim_{i\to\infty} c'_{n_i}$, and $e = \lim_{i\to\infty} e_{n_i}$, all exist. Then $F(e) = c$ by continuity. By construction, c differs from c' only at $1_\mathbb{G}$. By post-surjectivity, there exists a pre-image e' of c' such that $\Delta(e, e') \subseteq D_m$ for some $m \geq 0$. Take $\ell \gg m$ and choose k large enough such that $c'_{n_k}|_{D_\ell} = c'|_{D_\ell}$ and $e_{n_k}|_{D_\ell} = e|_{D_\ell}$. Define $\tilde{e}$ so that it agrees with e' on D_ℓ and with e_{n_k} outside D_m. Such $\tilde{e}$ does exist, because e', e, and e_{n_k} agree on $D_\ell \setminus D_m$. Then $\tilde{e}$ is a pre-image of c'_{n_k} which is asymptotic to e_{n_k} and agrees with e_{n_k} outside D_{n_k}, thus contradicting our assumption. □

By repeatedly applying Lemma 1 we get:

Corollary 1. *Let $\mathcal{A}$ be a post-surjective* CA *on a finitely generated group $\mathbb{G}$ and let F be its global transition function. There exists $N \geq 0$ such that, for every $r \geq 0$, however given three configurations c, c', e with $c = F(e)$ and $\Delta(c, c') \subseteq D_r$, there exists a pre-image e' of c' such that $\Delta(e, e') \subseteq D_{N+r}$.*

Assuming also pre-injectivity, we get the following stronger property:

Corollary 2. *Let $\mathcal{A}$ be a pre-injective, post-surjective* CA *on a finitely generated group $\mathbb{G}$ and let F be its global transition function. There exists $M \in \mathcal{PF}(\mathbb{G})$ with the following property: For every pair (e, e') of asymptotic configurations, if $c = F(e)$ and $c' = F(e')$ disagree at most on K, then e and e' disagree at most on KM.*

We are now ready to prove:

Theorem 1. *Every pre-injective, post-surjective cellular automaton on the full shift is reversible.*

Proof. By Proposition 4, it is sufficient to consider the case where $\mathbb{G}$ is finitely generated.

Let $\mathcal{A}$ be a pre-injective and post-surjective CA on the group $\mathbb{G}$, let S be its set of states, and let F be its global transition function. Let M be as in Corollary 2. We construct a new CA with neighborhood $\mathcal{N} = M^{-1}$. Calling H the global transition function of the new CA, we first prove that H is a *right* inverse of F. We then show that H is also a *left* inverse for F, thus completing the proof.

To construct the local update rule $h : S^{\mathcal{N}} \to S$, we proceed as follows. Fix a uniform configuration u and let $v = F(u)$. Given $g \in \mathbb{G}$ and $p : \mathcal{N} \to S$, for every $i \in \mathbb{G}$, put

$$
y_{g,p}(i) = \begin{cases} p(g^{-1}i) & \text{if } i \in g\mathcal{N} \\ v(i) & \text{otherwise} \end{cases} \tag{3}
$$

that is, let $y_{g,p}$ be obtained from v by cutting away the piece with support $g\mathcal{N}$ and pasting p as a "patch" for the "hole". By post-surjectivity and pre-injectivity combined, there exists a unique $x_{g,p} \in S^{\mathbb{G}}$ asymptotic to u such that $F(x_{g,p}) = y_{g,p}$. Let then

$$
h(p) = x_{g,p}(g). \tag{4}
$$

Observe that (4) does *not* depend on g: if $g' = i \cdot g$, then $y_{g',p} = \sigma_i(F(x_{g,p})) = F(\sigma_i(x_{g,p}))$, so that $x_{g',p} = \sigma_i(x_{g,p})$ by pre-injectivity, and $x_{g',p}(g') = x_{g,p}(g)$.

Let now y be *any* configuration asymptotic to v such that $y|_{g\mathcal{N}} = p$, and let x be the unique pre-image of y asymptotic to u. We claim that $x(g) = h(p)$. To prove this, we observe that, as y and $y_{g,p}$ are both asymptotic to v and they agree on $g\mathcal{N} = gM^{-1}$, the set K where they disagree is finite and is contained in $\mathbb{G} \setminus gM^{-1}$. By Corollary 2, their pre-images x and $x_{g,p}$ can disagree only on $KM \subseteq (\mathbb{G} \setminus gM^{-1}) M$. The set KM does not contain g, because if $g \in (\mathbb{G} \setminus gM^{-1}) M$, then for some $m \in M$, $gm^{-1} \in (\mathbb{G} \setminus gM^{-1})$, which is not the case! Therefore, $x(g) = x_{g,p}(g) = h(p)$, as we claimed.

The argument above holds whatever the pattern $p : \mathcal{N} \to S$ is. By applying it finitely many times to arbitrary finitely many points, we find the following fact: if y is any configuration which is asymptotic to v, then $F(H(y)) = y$. But the set of configurations asymptotic to v is dense in $S^{\mathbb{G}}$, so it follows from continuity of F and H that $F(H(y)) = y$ for every $y \in S^{\mathbb{G}}$.

We have thus shown that H is a right inverse of F. We next verify that H is also a left inverse of F.

Let x be a configuration asymptotic to u, and set $y = F(x)$. Note that y is asymptotic to v. The two configurations x and $H(y)$ are both asymptotic to u, and furthermore, $F(x) = y = F(H(y))$. Therefore, by the pre-injectivity of F, x and $H(y)$ must coincide, that is, $H(F(x)) = x$. The continuity of F and H now implies that the equality $H(F(x)) = x$ holds even if x is not asymptotic to u. Hence, H is a left inverse for F. □

Corollary 3. *A cellular automaton on an amenable group (in particular, a d-dimensional* CA*) is post-surjective if and only if it is reversible.*

4   Post-surjectivity on Sofic Groups

After proving Theorem 1, we might want to show examples of post-surjective cellular automata which are not pre-injective. However, the standard examples of surjective CA which are not pre-injective, such as the majority rule on the free group on two generators, fail to work. The reason is that, as we shall see below, finding such a counterexample amounts to finding a group which is not *sofic*, and that appears to be a difficult open problem.

The notion of sofic group was originally introduced by Gromov [9], but was later reformulated by Weiss [13] in combinatorial, rather than geometric, terms.

Definition 3. *Let* G *be a finitely generated group and let* B *be a finite symmetric set of generators for* G. *Let* $r \geq 0$ *be an integer and* $\varepsilon > 0$ *a real. An* (r, ε)-*approximation of* G *(relative to* B) *is a* B-*labeled graph* (V, E) *along with a subset* $U \subseteq V$ *such that the following hold:*

1. *For every* $u \in U$, *the neighborhood of radius* r *of* u *in* (V, E) *is isomorphic to* $D_{B,r}$ *as a labeled graph.*
2. $|U| > (1 - \varepsilon)|V|$.

The group G *is* sofic *(relative to* B) *if for every choice of* $r \geq 0$ *and* $\varepsilon > 0$, *there is an* (r, ε)-*approximation of* G *(relative to* B).

As explained in [13], the notion of soficness does not depend on the generating set B. For this reason, in the rest of this section, we will suppose B given once and for all. It is easy to see that finitely generated residually finite groups and finitely generated amenable groups are all sofic.

The importance of sofic groups is manifold: firstly, as per [13, Sect. 3], sofic groups are surjunctive; secondly, no examples of non-sofic groups are currently known. We add a third reason:

Theorem 2. *Let* G *be a sofic group. Every post-surjective cellular automaton on* G *is pre-injective.*

As a corollary, cellular automata which are post-surjective, but not pre-injective, could only exist over non-sofic groups!

To prove Theorem 2, we need two auxiliary lemmas. Observe that if $f : S^{D_R} \to S$ is the local rule of a cellular automaton on a group G with a finite generating set B, and (V, E) is a B-labeled graph, then f is applicable in an obvious fashion to patterns on V at every point $v \in V$ whose R-neighborhood in (V, E) is isomorphic to the disk of radius R in the Cayley graph of G with generating set B. Therefore, we extend our notation, and for two patterns $p : E \to S$ and $q : C \to S$ with $E, C \subseteq V$, we write $p \xrightarrow{f} q$ if for every $v \in C$, the R-neighborhood $D_R(v)$ is a subset of E and is isomorphic to the disk of radius R, and furthermore $f\big(p|_{D_R(v)} \big) = q(v)$.

Lemma 2. *Let $\mathcal{A}$ be a post-surjective* CA *on a sofic group $\mathbb{G}$. Let $\mathcal{A}$ have state set S, neighborhood radius R and local rule f, and let N be given by Lemma 1. Consider an (r, ε)-approximation given by a graph (V, E) and a set $U \subseteq V$, where $\varepsilon > 0$ and $r \geq N + 2R$. For every pattern $q : U \to S$, there is a pattern $p : V \to S$ such that $p \xrightarrow{f} q$.*

Proof. Take arbitrary $p_0 : V \to S$ and $q_0 : U \to S$ such that $p_0 \xrightarrow{f} q_0$. Let $q_0, q_1, \ldots, q_m = q$ be a sequence of patterns with support U such that, for every i, q_i and q_{i+1} only differ in a single $k_i \in U$. Since the r-neighborhood of k_i is isomorphic to the disk of the same radius from the Cayley graph of $\mathbb{G}$, we can apply Lemma 1 and deduce the existence of a sequence $p_0, p_1, \ldots, p_m$ with common support V such that each p_i is a pre-image of q_i and, for every i, p_i differs from p_{i+1} at most in $D_N(k_i)$. Then $p = p_m$ satisfies the thesis. $\square$

The next lemma is an observation made in [13].

Lemma 3 (Packing lemma). *Let $\mathbb{G}$ be a group with a finite generating set B. Let (V, E) be a B-labeled graph and $U \subseteq V$ a subset with $|U| \geq \frac{1}{2}|V|$ such that, for every $u \in U$, the 2ℓ-neighborhood of u in (V, E) is isomorphic to the disk of radius 2ℓ in the Cayley graph of $\mathbb{G}$. Then, there is a set $W \subseteq U$ of size at least $\frac{|V|}{2|D_{2\ell}|}$ such that the ℓ-neighborhoods of the elements of W are disjoint.*

Proof. Let $W \subseteq U$ be a maximal set such that the ℓ-neighborhoods of the elements of W are disjoint. Then, for every $u \in U$, the neighborhood $D_\ell(u)$ must intersect the set $\bigcup_{w \in W} D_\ell(w)$. Therefore, $U \subseteq D_{2\ell}(W)$, which gives $|U| \leq |D_{2\ell}| \cdot |W|$. $\square$

Proof (of Theorem 2). Let $\mathbb{G}$ be a sofic group and assume that $\mathcal{A} = \langle S, D_R, f \rangle$ is a cellular automaton on $\mathbb{G}$ that is post-surjective, but not pre-injective. For brevity, set $|S| = s \geq 2$. Let N be as in Lemma 1.

Since the CA is not pre-injective, there are two asymptotic configurations $x, x' : \mathbb{G} \to S$ such that $F_{\mathcal{A}}(x) = F_{\mathcal{A}}(x')$. Take m such that the disk D_m contains $\Delta(x, x')$. It follows that there are two mutually erasable patterns on D_{m+2R}, that is, two patterns $p, p' : D_{m+2R} \to S$ such that on any configuration z, replacing an occurrence of p with p' or vice versa does not change the image of z under $F_{\mathcal{A}}$.

Take $r \geq \max\{N, m\} + 2R$ and $\varepsilon > 0$ small. We shall need ε small enough so that

$$s^\varepsilon \cdot \left(1 - s^{-|D_r|}\right)^{\frac{1}{2|D_{2r}|}} < 1 \,.$$

Such a choice is possible, because the second factor on the left-hand side is a constant smaller than 1. Since $\mathbb{G}$ is sofic, there is a $(2r, \varepsilon)$-approximation of $\mathbb{G}$ given by a graph (V, E) and a set $U \subseteq V$. Let $\varphi : S^V \to S^U$ be the map given by $\varphi(p) = q$ if $p \xrightarrow{f} q$. Such φ is well defined, because the R-neighborhood of each $u \in U$ is isomorphic to the disk of radius R in $\mathbb{G}$.

By Lemma 2, the map φ is surjective, hence

$$|\varphi(S^V)| = s^{|U|} \,. \tag{5}$$

On the other hand, by Lemma 3, there is a collection $W \subseteq U$ of $|W| \geq \frac{|V|}{2|D_{2r}|}$ points in U whose r-neighborhoods are disjoint. Each of these r-neighborhoods is isomorphic to the disk $D_r \supseteq D_{m+2R}$ in $\mathbb{G}$. The existence of the mutually erasable patterns on D_r thus implies that there are at most

$$|\varphi(S^V)| \leq (s^{|D_r|} - 1)^{|W|} \cdot s^{|V| - |W| \cdot |D_r|}$$

patterns on V with distinct images. However,

$$(s^{|D_r|} - 1)^{|W|} \cdot s^{|V| - |W| \cdot |D_r|} = \left(1 - s^{-|D_r|}\right)^{|W|} \cdot s^{|V|}$$

$$\leq \left(1 - s^{-|D_r|}\right)^{\frac{|V|}{2|D_{2r}|}} \cdot s^{|V|}$$

$$< s^{-\varepsilon|V|} \cdot s^{|V|}$$

$$= s^{(1-\varepsilon)|V|}$$

$$< s^{|U|} :$$

which contradicts (5).   □

Corollary 4. *Let* $\mathbb{G}$ *be a sofic group and* $\mathcal{A}$ *a cellular automaton on* $\mathbb{G}$. *Then,* $\mathcal{A}$ *is post-surjective if and only if it is reversible.*

5   Conclusions

We have given a little contribution to a broad research theme by examining some links between different properties of cellular automata. In particular, we have seen how reversibility can still be obtained by weakening injectivity while strengthening surjectivity. Whether other such "transfers" are possible, is a field that we believe deserving to be explored. Another interesting issue is whether post-surjective cellular automata which are not pre-injective do or do not exist. By Theorem 2, such examples might exist only if non-sofic groups exist. We thus formulate the following "almost dual" to Gottschalk's conjecture:

Conjecture 2. Let $\mathbb{G}$ *be a group and* $\mathcal{A}$ *a cellular automaton on* $\mathbb{G}$. *If* $\mathcal{A}$ *is post-surjective, then it is pre-injective.*

References

1. Bartholdi, L.: Gardens of Eden and amenability on cellular automata. J. Eur. Math. Soc. **12**(1), 241–248 (2010)
2. Capobianco, S.: On the induction operation for shift subspaces and cellular automata as presentations of dynamical systems. Inform. Comput. **207**(11), 1169–1180 (2009)
3. Capobianco, S., Guillon, P., Kari, J.: Surjective cellular automata far from the Garden of Eden. Disc. Math. Theor. Comput. Sci. **15**(3), 41–60 (2013)

4. Capobianco, S., Kari, J., Taati, S.: Post-surjectivity and balancedness of cellular automata over groups. In: Kari, J., Törmä, I., Szabados, M. (eds) 21st International Workshop on Cellular Automata and Discrete Complex Systems: Exploratory Papers of AUTOMATA 2015, Turku Centre for Computer Science, TUCS Lecture Notes, vol. 24, pp. 31–38 (2015)

5. Ceccherini-Silberstein, T., Coornaert, M.: Cellular Automata and Groups. Springer Monographs in Mathematics. Springer, Heidelberg (2010)

6. Ceccherini-Silberstein, T., Machì, A., Scarabotti, F.: Amenable groups and cellular automata. Ann. Inst. Fourier **49**(2), 673–685 (1999)

7. Fiorenzi, F.: Cellular automata and strongly irreducible shifts of finite type. Theor. Comput. Sci. **299**, 477–493 (2003)

8. Gottschalk, W.H.: Some general dynamical notions. In: Gottschalk, W. (ed.) Recent Advances in Topological Dynamics. Lecture Notes in Mathematics, vol. 318. Springer, Heidelberg (1973)

9. Gromov, M.: Endomorphisms of symbolic algebraic varieties. J. European Math. Soc. **1**, 109–197 (1999)

10. Kari, J.: Theory of cellular automata: a survey. Theor. Comp. Sci. **334**, 3–33 (2005)

11. Kari, J., Taati, S.: Statistical mechanics of surjective cellular automata. J. Stat. Phys. **160**(5), 1198–1243 (2015)

12. Lind, D., Marcus, B.: An Introduction to Symbolic Dynamics and Coding. Cambridge University Press, Cambridge (1995)

13. Weiss, B.: Sofic groups and dynamical systems. Sankhyā: Indian. J Stat. **62A**(3), 350–359 (2000)

On Finite Monoids of Cellular Automata

Alonso Castillo-Ramirez$^{(\boxtimes)}$ and Maximilien Gadouleau

School of Engineering and Computing Sciences, Durham University, South Road,
Durham DH1 3LE, UK
{alonso.castillo-ramirez,m.r.gadouleau}@durham.ac.uk

Abstract. For any group G and set A, a cellular automaton over G and
A is a transformation $\tau : A^G \to A^G$ defined via a finite neighbourhood
$S \subseteq G$ (called a memory set of τ) and a local function $\mu : A^S \to A$.
In this paper, we assume that G and A are both finite and study various
algebraic properties of the finite monoid $\mathrm{CA}(G, A)$ consisting of all cellular automata over G and A. Let $\mathrm{ICA}(G; A)$ be the group of invertible
cellular automata over G and A. In the first part, using information on
the conjugacy classes of subgroups of G, we give a detailed description
of the structure of $\mathrm{ICA}(G; A)$ in terms of direct and wreath products.
In the second part, we study generating sets of $\mathrm{CA}(G; A)$. In particular,
we prove that $\mathrm{CA}(G, A)$ cannot be generated by cellular automata with
small memory set, and, when G is finite abelian, we determine the mini-
mal size of a set $V \subseteq \mathrm{CA}(G; A)$ such that $\mathrm{CA}(G; A) = \langle \mathrm{ICA}(G; A) \cup V \rangle$.

Keywords: Cellular automata · Invertible cellular automata ·
Monoids · Generating sets

1 Introduction

Cellular automata (CA), first introduced by John von Neumann as an attempt
to design self-reproducing systems, are models of computation with important
applications to computer science, physics, and theoretical biology. In recent
years, the theory of CA has been greatly enriched with its connections to group
theory and topology (see [4] and references therein). One of the goals of this
paper is to embark in the new task of exploring CA from the point of view of
finite group and semigroup theory.

We review the broad definition of CA that appears in [4, Sect. 1.4]. Let G
be a group and A a set. Denote by A^G the *configuration space*, i.e. the set of all
functions of the form $x : G \to A$. For each $g \in G$, let $R_g : G \to G$ be the right
multiplication function, i.e. $(h)R_g := hg$, for any $h \in G$. We emphasise that we
apply functions on the right, while in [4] functions are applied on the left.

Definition 1. Let G be a group and A a set. A *cellular automaton* over G and
A is a transformation $\tau : A^G \to A^G$ such that there is a finite subset $S \subseteq G$,
called a *memory set* of τ, and a *local function* $\mu : A^S \to A$ satisfying

© IFIP International Federation for Information Processing 2016
Published by Springer International Publishing Switzerland 2016. All Rights Reserved
M. Cook and T. Neary (Eds.): AUTOMATA 2016, LNCS 9664, pp. 90–104, 2016.
DOI: 10.1007/978-3-319-39300-1_8

$$(g)(x)\tau = ((R_g \circ x)|_S)\mu, \ \forall x \in A^G, g \in G.$$

Most of the classical literature on CA focuses on the case when $G = \mathbb{Z}^d$, for $d \geq 1$, and A is a finite set (e.g. see survey [11]).

A *semigroup* is a set M equipped with an associative binary operation. If there exists an element id $\in M$ such that id$\cdot m = m \cdot$id $= m$, for all $m \in M$, the semigroup M is called a *monoid* and id an *identity* of M. Clearly, the identity of a monoid is always unique.

Let $CA(G; A)$ be the set of all cellular automata over G and A; by [4, Corollary 1.4.11], this set equipped with the composition of functions is a monoid. Although results on monoids of CA have appeared in the literature before (see [3,9,12]), the algebraic structure of $CA(G; A)$ remains basically unknown. In particular, the study of $CA(G; A)$, when G and A are both finite, has been generally disregarded, perhaps because some of the classical questions are trivially answered (e.g. the Garden of Eden theorems become trivial). However, many new questions, typical of finite semigroup theory, arise in this setting.

In this paper, we study various algebraic properties of $CA(G; A)$ when G and A are both finite. First, in Sect. 2, we introduce notation and review some basic results. In Sect. 3, we study the group $ICA(G; A)$ consisting of all invertible CA: we show that its structure is linked with the number of conjugacy classes of subgroups of G, and we give an explicit decomposition in terms of direct and wreath products.

In Sect. 4, we study generating sets of $CA(G; A)$. We prove that $CA(G; A)$ cannot be generated by CA with small memory sets: if T generates $CA(G; A)$, then T must contain a cellular automaton with minimal memory set equal to G itself. This result provides a striking contrast with CA over infinite groups. Finally, when G is finite abelian, we find the smallest size of a set $U \subseteq CA(G; A)$ such that $ICA(G; A) \cup U$ generate $CA(G; A)$; this number is known in semigroup theory as the *relative rank* of $ICA(G; A)$ in $CA(G; A)$, and it turns out to be related with the number of edges of the subgroup lattice of G.

2 Basic Results

For any set X, let $\mathrm{Tran}(X)$ and $\mathrm{Sym}(X)$ be the sets of all functions and bijective functions, respectively, of the form $\tau : X \to X$. Equipped with the composition of functions, $\mathrm{Tran}(X)$ is known as the *full transformation monoid* on X, while $\mathrm{Sym}(X)$ is the *symmetric group* on X. When X is finite and $|X| = q$, we write Tran_q and Sym_q instead of $\mathrm{Tran}(X)$ and $\mathrm{Sym}(X)$, respectively.

A *finite transformation monoid* is simply a submonoid of Tran_q, for some q. This type of monoids has been extensively studied (e.g. see [6] and references therein), and it should be noted its close relation to finite-state machines.

For the rest of the paper, let G be a finite group of size n and A a finite set of size q. By Definition 1, it is clear that $CA(G; A) \leq \mathrm{Tran}(A^G)$ (we use the symbol "$\leq$" for the submonoid relation). We may always assume that $\tau \in CA(G; A)$ has (not necessarily minimal) memory set $S = G$, so τ is completely determined by its local function $\mu : A^G \to A$. Hence, $|CA(G; A)| = q^{q^n}$.

If $n = 1$, then $\mathrm{CA}(G; A) = \mathrm{Tran}(A)$, while, if $q \leq 1$, then $\mathrm{CA}(G; A)$ is the trivial monoid with one element; henceforth, we assume $n \geq 2$ and $q \geq 2$. We usually identify A with the set $\{0, 1, \ldots, q - 1\}$.

The group G acts on the configuration space A^G as follows: for each $g \in G$ and $x \in A^G$, the configuration $x \cdot g \in A^G$ is defined by

$$(h)x \cdot g = (hg^{-1})x, \quad \forall h \in G.$$

A transformation $\tau : A^G \to A^G$ is G-equivariant if, for all $x \in A^G$, $g \in G$,

$$(x \cdot g)\tau = ((x)\tau) \cdot g.$$

Denote by $\mathrm{ICA}(G; A)$ the group of all invertible cellular automata:

$$\mathrm{ICA}(G; A) := \{\tau \in \mathrm{CA}(G; A) : \exists \phi \in \mathrm{CA}(G; A) \text{ such that } \tau\phi = \phi\tau = \mathrm{id}\}.$$

Theorem 1. *Let G be a finite group and A a finite set.*

(i) $\mathrm{CA}(G; A) = \{\tau \in \mathrm{Tran}(A^G) : \tau \text{ is } G\text{-equivariant}\}$.
(ii) $\mathrm{ICA}(G; A) = \mathrm{CA}(G; A) \cap \mathrm{Sym}(A^G)$.

Proof. The first part follows by Curtis-Hedlund Theorem (see [4, Theorem 1.8.1]) while the second part follows by [4, Theorem 1.10.2]. $\quad\square$

Notation 1. *For any $x \in A^G$, denote by xG the G-orbit of x on A^G:*

$$xG := \{x \cdot g : g \in G\}.$$

Let $\mathcal{O}(G; A)$ be the set of all G-orbits on A^G:

$$\mathcal{O}(G; A) := \{xG : x \in A^G\}.$$

Clearly, $\mathcal{O}(G; A)$ forms a partition of A^G. In general, when X is a set and $\mathcal{P}$ is a partition of X, we say that a transformation monoid $M \leq \mathrm{Tran}(X)$ *preserves the partition* if, for any $P \in \mathcal{P}$ and $\tau \in M$ there is $Q \in \mathcal{P}$ such that $(P)\tau \subseteq Q$.

Lemma 1. *For any $x \in A^G$ and $\tau \in \mathrm{CA}(G; A)$,*

$$(xG)\tau = (x)\tau G.$$

In particular, $\mathrm{CA}(G; A)$ preserves the partition $\mathcal{O}(G; A)$ of A^G.

Proof. The result follows by the G-equivariance of $\tau \in \mathrm{CA}(G; A)$. $\quad\square$

A configuration $x \in A^G$ is called *constant* if $(g)x = k \in A$, for all $g \in G$. In such case, we usually denote x by $\mathbf{k} \in A^G$.

Lemma 2. *Let $\tau \in \mathrm{CA}(G; A)$ and let $\mathbf{k} \in A^G$ be a constant configuration. Then, $(\mathbf{k})\tau \in A^G$ is a constant configuration.*

Proof. Observe that $x \in A^G$ is constant if and only if $x \cdot g = x$, for all $g \in G$. By G-equivariance,

$$(\mathbf{k})\tau = (\mathbf{k} \cdot g)\tau = (\mathbf{k})\tau \cdot g, \quad \forall g \in G.$$

Hence, $(\mathbf{k})\tau$ is constant. $\square$

For a monoid M and a subset $T \subseteq M$, denote by $C_M(T)$ the *centraliser* of T in M:

$$C_M(T) := \{m \in M : mt = tm, \forall t \in T\}.$$

If G is abelian, the transformation $\sigma_g : A^G \to A^G$, with $g \in G$, defined by

$$(x)\sigma_g := x \cdot g, \quad \forall x \in A^G,$$

is in $CA(G; A)$. It follows by Theorem 1 that $CA(G; A) = C_{\text{Tran}(A^G)}(T)$, where $T := \{\sigma_g : g \in G\}$.

We use the cyclic notation for the permutations of $\text{Sym}(A^G)$. If $B \subseteq A^G$ and $a \in A^G$, we define the idempotent transformation $(B \to a) \in \text{Tran}(A^G)$ by

$$(x)(B \to a) := \begin{cases} a & \text{if } x \in B, \\ x & \text{otherwise}, \end{cases} \quad \forall x \in A^G.$$

When $B = \{b\}$ is a singleton, we write $(b \to a)$ instead of $(\{b\} \to a)$.

3   The Structure of ICA$(G; A)$

Let G be a finite group of size $n \geq 2$ and A a finite set of size $q \geq 2$. We review few basic concepts about permutation groups (see [5, Ch.1]). For $x \in A^G$, denote by G_x the *stabiliser* of x in G:

$$G_x := \{g \in G : x \cdot g = x\}.$$

Remark 1. For any subgroup $H \leq G$ there exists $x \in A^G$ such that $G_x = H$; namely, we may define $x : G \to A$ by

$$(g)x := \begin{cases} 1 & \text{if } g \in H, \\ 0 & \text{otherwise}, \end{cases} \quad \forall g \in G.$$

Say that two subgroups H_1 and H_2 of G are *conjugate* in G if there exists $g \in G$ such that $g^{-1}H_1 g = H_2$. This defines an equivalence relation on the subgroups of G. Denote by $[H]$ the conjugacy class of $H \leq G$.

We say that the actions of G on two sets Ω and Γ are *equivalent* if there is a bijection $\lambda : \Omega \to \Gamma$ such that, for all $x \in \Omega, g \in G$, we have $(x \cdot g)\lambda = (x)\lambda \cdot g$.

The following is an essential result for our description of the structure of the group of invertible cellular automata.

Lemma 3. *Let G be a finite group of size $n \geq 2$ and A a finite set of size $q \geq 2$. For any $x, y \in A^G$, there exists $\tau \in \mathrm{ICA}(G; A)$ such that $(xG)\tau = yG$ if and only if $[G_x] = [G_y]$.*

Proof. By [5, Lemma 1.6B], the actions of G on xG and yG are equivalent if and only if G_x and G_y are conjugate in G. We claim that the actions of G on xG and yG are equivalent if and only if there is $\tau \in \mathrm{ICA}(G; A)$ such that $(xG)\tau = yG$. Assume such $\tau \in \mathrm{ICA}(G; A)$ exists. Then, the restriction $\lambda := \tau|_{xG} : xG \to yG$ is the bijection required to show that the actions of G on xG and yG are equivalent. Conversely, suppose there is a bijection $\lambda : xG \to yG$ such that $(z \cdot g)\lambda = (z)\lambda \cdot g$, for all $z \in xG$, $g \in G$. Define $\tau : A^G \to A^G$ by

$$(z)\tau := \begin{cases} (z)\lambda & \text{if } z \in xG, \\ (z)\lambda^{-1} & \text{if } z \in yG, \quad \forall z \in A^G. \\ z & \text{otherwise,} \end{cases}$$

Clearly, τ is G-equivariant and invertible (in fact, $\tau = \tau^{-1}$). Hence $\tau \in \mathrm{ICA}(G; A)$, and it satisfies $(xG)\tau = yG$. □

Corollary 1. *Suppose that G is a finite abelian group. For any $x, y \in A^G$, there exists $\tau \in \mathrm{ICA}(G; A)$ such that $(xG)\tau = yG$ if and only if $G_x = G_y$.*

For any integer $\alpha \geq 2$ and any group C, the *wreath product* of C by Sym_α is the set

$$C \wr \mathrm{Sym}_\alpha := \{(v; \phi) : v \in C^\alpha, \phi \in \mathrm{Sym}_\alpha\}$$

equipped with the operation

$$(v; \phi) \cdot (w; \psi) = (vw^\phi; \phi\psi), \text{ for any } v, w \in C^\alpha, \phi, \psi \in \mathrm{Sym}_\alpha,$$

where ϕ acts on w by permuting its coordinates:

$$w^\phi = (w_1, w_2, \ldots, w_\alpha)^\phi := (w_{(1)\phi}, w_{(2)\phi}, \ldots, w_{(\alpha)\phi}).$$

See [5, Sect. 2.6] for a more detailed description of the wreath product.

Notation 2. *Let $O \in \mathcal{O}(G; A)$ be a G-orbit. If $G_{(O)}$ is the pointwise stabiliser of O, i.e. $G_{(O)} := \bigcap_{x \in O} G_x$, the group $G^O := G/G_{(O)}$ is isomorphic to a subgroup of $\mathrm{Sym}(O)$ (see [5, p. 17]). Abusing the notation, we also write G^O for the isomorphic copy of G^O inside $\mathrm{Sym}(O)$. Define the group*

$$C(G^O) := \{\tau|_O : O \to A^G : \tau \in \mathrm{ICA}(G; A) \text{ and } (O)\tau = O\}. \tag{1}$$

By Theorem 1, $C(G^O)$ is isomorphic to the centraliser of G^O in $\mathrm{Sym}(O)$, i.e. $C(G^O) \cong C_{\mathrm{Sym}(O)}(G^O)$.

Notation 3. *Let H be a subgroup of G and $[H]$ its conjugacy class. Define*

$$B_{[H]} := \{x \in A^G : G_x \in [H]\}.$$

Note that $B_{[H]}$ is a union of G-orbits and, by the Orbit-Stabiliser Theorem (see [5, Theorem 1.4A]), all the G-orbits contained in $B_{[H]}$ have equal sizes. Define

$$\alpha_{[H]}(G; A) := \left| \{ O \in \mathcal{O}(G, A) : O \subseteq B_{[H]} \} \right|.$$

If r is the number of different conjugacy classes of subgroups of G, observe that

$$\mathcal{B} := \{ B_{[H]} : H \leq G \}$$

is a partition of A^G with r blocks.

Remark 2. $B_{[G]} = \{ x \in A^G : x \text{ is constant} \}$ and $\alpha_{[G]}(G; A) = q$.

Example 1. Let $G = \mathbb{Z}_2 \times \mathbb{Z}_2$ be the Klein four-group and $A = \{0, 1\}$. As G is abelian, $[H] = \{H\}$, for all $H \leq G$. The subgroups of G are

$$H_1 = G, \ H_2 = \langle (1,0) \rangle, \ H_3 = \langle (0,1) \rangle, \ H_4 = \langle (1,1) \rangle, \text{ and } H_5 = \langle (0,0) \rangle,$$

where $\langle (a,b) \rangle$ denotes the subgroup generated by $(a,b) \in G$. Any configuration $x : G \to A$ may be written as a 2×2 matrix $(x_{i,j})$ where $x_{i,j} := (i-1, j-1)x$, $i, j \in \{1, 2\}$. The G-orbits on A^G are

$$O_1 := \left\{ \begin{pmatrix} 0 & 0 \\ 0 & 0 \end{pmatrix} \right\}, \ O_2 := \left\{ \begin{pmatrix} 1 & 1 \\ 1 & 1 \end{pmatrix} \right\}, \ O_3 := \left\{ \begin{pmatrix} 1 & 0 \\ 1 & 0 \end{pmatrix}, \begin{pmatrix} 0 & 1 \\ 0 & 1 \end{pmatrix} \right\},$$

$$O_4 := \left\{ \begin{pmatrix} 1 & 1 \\ 0 & 0 \end{pmatrix}, \begin{pmatrix} 0 & 0 \\ 1 & 1 \end{pmatrix} \right\}, \ O_5 := \left\{ \begin{pmatrix} 1 & 0 \\ 0 & 1 \end{pmatrix}, \begin{pmatrix} 0 & 1 \\ 1 & 0 \end{pmatrix} \right\}$$

$$O_6 := \left\{ \begin{pmatrix} 1 & 0 \\ 0 & 0 \end{pmatrix}, \begin{pmatrix} 0 & 1 \\ 0 & 0 \end{pmatrix}, \begin{pmatrix} 0 & 0 \\ 0 & 1 \end{pmatrix}, \begin{pmatrix} 0 & 0 \\ 1 & 0 \end{pmatrix} \right\},$$

$$O_7 := \left\{ \begin{pmatrix} 0 & 1 \\ 1 & 1 \end{pmatrix}, \begin{pmatrix} 1 & 0 \\ 1 & 1 \end{pmatrix}, \begin{pmatrix} 1 & 1 \\ 1 & 0 \end{pmatrix}, \begin{pmatrix} 1 & 1 \\ 0 & 1 \end{pmatrix} \right\}.$$

Hence,

$$B_{[H_1]} := O_1 \cup O_2, \ B_{[H_2]} := O_3, \ B_{[H_3]} := O_4, \ B_{[H_4]} := O_5, \ B_{[H_5]} := O_6 \cup O_7;$$
$$\alpha_{[H_i]}(G; A) = 2, \text{ for } i \in \{1, 5\}, \text{ and } \alpha_{[H_i]}(G; A) = 1, \text{ for } i \in \{2, 3, 4\}.$$

Remark 3. By Lemma 3, the $\mathrm{ICA}(G; A)$-orbits on A^G coincide with the blocks in $\mathcal{B}$, while the $\mathrm{ICA}(G; A)$-blocks of imprimitivity on each $B_{[H]}$ are the G-orbits contained in $B_{[H]}$.

The following result is a refinement of [12, Theorem 9] and [3, Lemma 4].

Theorem 2. *Let G be a finite group and A a finite set of size $q \geq 2$. Let $[H_1], \dots, [H_r]$ be the list of different conjugacy classes of subgroups of G. For each $1 \leq i \leq r$, fix a G-orbit $O_i \subseteq B_{[H_i]}$. Then,*

$$\mathrm{ICA}(G; A) \cong \prod_{i=1}^{r} \left(C_i \wr \mathrm{Sym}_{\alpha_i} \right),$$

where $C_i := C(G^{O_i}) \cong C_{\mathrm{Sym}(O_i)}(G^{O_i})$ and $\alpha_i := \alpha_{[H_i]}(G; A)$.

Proof. Let $B_i := B_{[H_i]}$. By Lemma 3, ICA$(G; A)$ is contained in the group

$$\prod_{i=1}^{r} \text{Sym}(B_i) = \text{Sym}(B_1) \times \text{Sym}(B_2) \times \cdots \times \text{Sym}(B_r).$$

For each $1 \le i \le r$, let $\mathcal{O}_i$ be the set of G-orbits contained in B_i (so $O_i \in \mathcal{O}_i$). Note that $\mathcal{O}_i$ is a uniform partition of B_i. For any $\tau \in$ ICA$(G; A)$, Lemma 1 implies that the projection of τ to $\text{Sym}(B_i)$ is contained in

$$S(B_i, \mathcal{O}_i) := \{\phi \in \text{Sym}(B_i) : \forall P \in \mathcal{O}_i, \ (P)\phi \in \mathcal{O}_i\}.$$

By [2, Lemma 2.1(iv)],

$$S(B_i, \mathcal{O}_i) \cong \text{Sym}(\mathcal{O}_i) \wr \text{Sym}_{\alpha_i}.$$

It is well-known that Sym_{α_i} is generated by its transpositions. As the invertible cellular automaton constructed in the proof of Lemma 3 induces a transposition $(xG, yG) \in \text{Sym}_{\alpha_i}$, with $xG, yG \in \mathcal{O}_i$, we deduce that $\text{Sym}_{\alpha_i} \le$ ICA$(G; A)$. The result follows by the construction of $C_i \cong C_{\text{Sym}(O_i)}(G^{O_i})$ and Theorem 1. $\qquad\square$

Corollary 2. *Let G be a finite abelian group and A a finite set of size $q \ge 2$. Let $H_1, \ldots, H_r$ be the list of different subgroups of G. Then,*

$$\text{ICA}(G; A) \cong \prod_{i=1}^{r} \left((G/H_i) \wr \text{Sym}_{\alpha_i} \right),$$

and $|G|\alpha_i = |H_i| \cdot |\{x \in A^G : G_x = H_i\}|$, where $\alpha_i := \alpha_{[H_i]}(G; A)$.

Proof. By [5, Theorem 4.2A(v)], $C_{\text{Sym}(O_i)}(G^{O_i}) \cong G^{O_i} \cong G/G_{x_i}$, where $x_i \in O_i$. By Remark 1, the list of pointwise stabilisers coincide with the list of subgroups of G, and, as G is abelian, $[H_i] = \{H_i\}$ for all i. Finally, by the Orbit-Stabiliser theorem, every orbit contained in $B_i = \{x \in A^G : G_x = H_i\}$ has size $\frac{|G|}{|H_i|}$; as these orbits form a partition of B_i, we have $|B_i| = \alpha_i \frac{|G|}{|H_i|}$. $\qquad\square$

Example 2. Let $G = \mathbb{Z}_2 \times \mathbb{Z}_2$ and $A = \{0, 1\}$. By Example 1,

$$\text{ICA}(G, A) \cong (\mathbb{Z}_2)^4 \times (G \wr \text{Sym}_2).$$

4   Generating Sets of of CA$(G; A)$

For a monoid M and a subset $T \subseteq M$, denote by $\langle T \rangle$ the submonoid *generated* by T, i.e. the smallest submonoid of M containing T. Say that T is a *generating set* of M if $M = \langle T \rangle$; in this case, every element of M is expressible as a word in the elements of T (we use the convention that the empty word is the identity).

Define the *kernel* of a transformation $\tau : X \to X$, denoted by $\ker(\tau)$, as set of the equivalence classes on X of the equivalence relation $\{(x, y) \in X^2 : (x)\tau = (y)\tau\}$. For example, $\ker(\phi) = \{\{x\} : x \in X\}$, for any $\phi \in \mathrm{Sym}(X)$, while $\ker(y \to z) = \{\{y, z\}, \{x\} : x \in X \setminus \{y, z\}\}$, for $y, z \in X$, $y \neq z$.

A large part of the classical research on CA has been focused on CA with small memory sets. In some cases, such as the elementary Rule 110, or John Conway's Game of Life, these CA are known to be Turing complete. In a striking contrast, when G and A are both finite, CA with small memory sets are insufficient to generate the monoid $\mathrm{CA}(G; A)$.

Theorem 3. *Let G be a finite group of size $n \geq 2$ and A a finite set of size $q \geq 2$. Let T be a generating set of $\mathrm{CA}(G; A)$. Then, there exists $\tau \in T$ with minimal memory set $S = G$.*

Proof. Suppose that T is a generating set of $\mathrm{CA}(G, A)$ such that each of its elements has minimal memory set of size at most $n-1$. Consider the idempotent $\sigma := (\mathbf{0} \to \mathbf{1}) \in \mathrm{CA}(G, A)$, where $\mathbf{0}, \mathbf{1} \in A^G$ are different constant configurations. Then, $\sigma = \tau_1 \tau_2 \ldots \tau_\ell$, for some $\tau_i \in T$. By the definition of σ and Lemma 2, there must be $1 \leq j \leq \ell$ such that $\left|(A_c^G)\tau_j\right| = q - 1$ and $(A_{nc}^G)\tau_j = A_{nc}^G$, where

$$A_c^G := \{\mathbf{k} \in A^G : \mathbf{k} \text{ is constant}\} \text{ and } A_{nc}^G := \{x \in A^G : x \text{ is non-constant}\}.$$

Let $S \subseteq G$ and $\mu : A^S \to A$ be the minimal memory set and local function of $\tau := \tau_j$, respectively. By hypothesis, $s := |S| < n$. Since the restriction of τ to A_c^G is not a bijection, there exists $\mathbf{k} \in A_c^G$ (defined by $(g)\mathbf{k} := k \in A$, $\forall g \in G$) such that $\mathbf{k} \notin (A_c^G)\tau$.

For any $x \in A^G$, define the *k-weight* of x by

$$|x|_k := |\{g \in G : (g)x \neq k\}|.$$

Consider the sum of the k-weights of all non-constant configurations of A^G:

$$w := \sum_{x \in A_{nc}^G} |x|_k = n(q-1)q^{n-1} - n(q-1) = n(q-1)(q^{n-1} - 1).$$

In particular, $\frac{w}{n}$ is an integer not divisible by q.

For any $x \in A^G$ and $y \in A^S$, define

$$\mathrm{Sub}(y, x) := |\{g \in G : y = x|_{Sg}\}|;$$

this counts the number of times that y appears as a subconfiguration of x. Then, for any fixed $y \in A^S$,

$$N_y := \sum_{x \in A_{nc}^G} \mathrm{Sub}(y, x) = \begin{cases} nq^{n-s} & \text{if } y \in A_{nc}^S, \\ n(q^{n-s} - 1) & \text{if } y \in A_c^S. \end{cases}$$

Let $\delta : A^2 \to \{0,1\}$ be the Kronecker's delta function. Since $(A_{nc}^G)\tau = A_{nc}^G$, we have

$$w = \sum_{x \in A_{nc}^G} |(x)\tau|_k = \sum_{y \in A^S} N_y(1 - \delta((y)\mu, k))$$

$$= nq^{n-s} \sum_{y \in A_{nc}^S} (1 - \delta((y)\mu, k)) + n(q^{n-s} - 1) \sum_{y \in A_c^S} (1 - \delta((y)\mu, k)).$$

Because $\mathbf{k} \notin (A_c^G)\tau$, we know that $(y)\mu \neq k$ for all $y \in A_c^S$. Therefore,

$$\frac{w}{n} = q^{n-s} \sum_{y \in A_{nc}^S} (1 - \delta_{(y)\mu, k}) + (q^{n-s} - 1)q.$$

As $s < n$, this implies that $\frac{w}{n}$ is an integer divisible by q, which is a contradiction. $\qquad\square$

One of the fundamental problems in the study of a finite monoid M is the determination of the cardinality of a smallest generating subset of M; this is called the *rank* of M and denoted by $\mathrm{Rank}(M)$:

$$\mathrm{Rank}(M) := \min\{|T| : T \subseteq M \text{ and } \langle T \rangle = M\}.$$

It is well-known that, if X is any finite set, the rank of the full transformation monoid $\mathrm{Tran}(X)$ is 3, while the rank of the symmetric group $\mathrm{Sym}(X)$ is 2 (see [6, Ch. 3]). Ranks of various finite monoids have been determined in the literature before (e.g. see [1,2,7,8,10]).

In [3], the rank of $\mathrm{CA}(\mathbb{Z}_n, A)$, where $\mathbb{Z}_n$ is the cyclic group of order n, was studied and determined when $n \in \{p, 2^k, 2^k p : k \geq 1,\ p \text{ odd prime}\}$. Moreover, the following problem was proposed:

Problem 1. For any finite group G and finite set A, determine $\mathrm{Rank}(\mathrm{CA}(G; A))$.

For any finite monoid M and $U \subseteq M$, the *relative rank* of U in M, denoted by $\mathrm{Rank}(M : U)$, is the minimum cardinality of a subset $V \subseteq M$ such that $\langle U \cup V \rangle = M$. For example, for any finite set X,

$$\mathrm{Rank}(\mathrm{Tran}(X) : \mathrm{Sym}(X)) = 1,$$

as any $\tau \in \mathrm{Tran}(X)$ with $|(X)\tau| = |X| - 1$ satisfies $\langle \mathrm{Sym}(X) \cup \{\tau\} \rangle = \mathrm{Tran}(X)$. One of the main tools that may be used to determine $\mathrm{Rank}(\mathrm{CA}(G; A))$ is based on the following result (see [2, Lemma 3.1]).

Lemma 4. *Let G be a finite group and A a finite set. Then,*

$$\mathrm{Rank}(\mathrm{CA}(G; A)) = \mathrm{Rank}(\mathrm{CA}(G; A) : \mathrm{ICA}(G; A)) + \mathrm{Rank}(\mathrm{ICA}(G; A)).$$

We shall determine the relative rank of $\mathrm{ICA}(G; A)$ in $\mathrm{CA}(G; A)$ for any finite abelian group G and finite set A. In order to achieve this, we prove two lemmas that hold even when G is nonabelian and have relevance in their own right.

Lemma 5. *Let G be a finite group and A a finite set of size $q \geq 2$. Let $x \in A^G$. If $(x)\tau \in xG$, then $\tau|_{xG} \in \mathrm{Sym}(xG)$.*

Proof. It is enough to show that $(xG)\tau = xG$ as this implies that $\tau|_{xG} : xG \to xG$ is surjective, so it is bijective by the finiteness of xG. Since $(x)\tau \in xG$, we know that $(x)\tau G = xG$. Hence, by Lemma 1,

$$(xG)\tau = (x)\tau G = xG. \qquad \square$$

Notation 4. *Denote by $\mathcal{C}_G$ the set of conjugacy classes of subgroups of G. For any $[H_1], [H_2] \in \mathcal{C}_G$, write $[H_1] \leq [H_2]$ if $H_1 \leq g^{-1}H_2g$, for some $g \in G$.*

Remark 4. The relation $\leq$ defined above is a well-defined partial order on $\mathcal{C}_G$. Clearly, $\leq$ is reflexive and transitive. In order to show antisymmetry, suppose that $[H_1] \leq [H_2]$ and $[H_2] \leq [H_1]$. Then, $H_1 \leq g^{-1}H_2g$ and $H_2 \leq f^{-1}H_1f$, for some $f, g \in G$, which implies that $|H_1| \leq |H_2|$ and $|H_2| \leq |H_1|$. As H_1 and H_2 are finite, $|H_1| = |H_2|$, and $H_1 = g^{-1}H_2g$. This shows that $[H_1] = [H_2]$.

Lemma 6. *Let G be a finite group and A a finite set of size $q \geq 2$. Let $x, y \in A^G$ be such that $xG \neq yG$. There exists a non-invertible $\tau \in \mathrm{CA}(G; A)$ such that $(xG)\tau = yG$ if and only if $[G_x] \leq [G_y]$.*

Proof. Suppose that $[G_x] \leq [G_y]$. Then, $G_x \leq g^{-1}G_yg$, for some $g \in G$. We define an idempotent $\tau_{x,y} : A^G \to A^G$ that maps xG to yG:

$$(z)\tau_{x,y} := \begin{cases} y \cdot gh & \text{if } z = x \cdot h, \\ z & \text{otherwise,} \end{cases} \quad \forall z \in A^G.$$

We verify that $\tau_{x,y}$ is well-defined. If $x \cdot h_1 = x \cdot h_2$, for $h_i \in G$, then $h_1 h_2^{-1} \in G_x$. As $G_x \leq g^{-1}G_yg$, we have $h_1 h_2^{-1} = g^{-1}sg$, for some $s \in G_y$. Thus, $gh_1 = sgh_2$ implies that $y \cdot gh_1 = y \cdot gh_2$, and $(x \cdot h_1)\tau_{x,y} = (x \cdot h_2)\tau_{x,y}$. Clearly, $\tau_{x,y}$ is non-invertible and G-equivariant, so $\tau_{x,y} \in \mathrm{CA}(G; A)$.

Conversely, suppose there exists $\tau \in \mathrm{CA}(G; A)$ such that $(xG)\tau = yG$. Then, $(x)\tau = y \cdot h$, for some $h \in G$. Let $s \in G_x$. By G-equivariance,

$$y \cdot h = (x)\tau = (x \cdot s)\tau = (x)\tau \cdot s = y \cdot hs.$$

Thus $hsh^{-1} \in G_y$ and $s \in h^{-1}G_yh$. This shows that $[G_x] \leq [G_y]$. $\qquad \square$

Corollary 3. *Suppose that G is finite abelian. Let $x, y \in A^G$ be such that $xG \neq yG$. There exists $\tau_{x,y} \in \mathrm{CA}(G; A)$ such that $(x)\tau_{x,y} = y$ and $(z)\tau_{x,y} = z$ for all $z \in A^G \setminus xG$ if and only if $G_x \leq G_y$.*

Notation 5. *Consider the directed graph $(\mathcal{C}_G, \mathcal{E}_G)$ with vertex set $\mathcal{C}_G$ and edge set*

$$\mathcal{E}_G := \{([H_i], [H_j]) \in \mathcal{C}_G^2 : [H_i] \leq [H_j]\}.$$

When G is abelian, this graph coincides with the lattice of subgroups of G.

Remark 5. Lemma 6 may be restated in terms of $\mathcal{E}_G$. By Lemma 5, loops $([H_i], [H_i])$ do not have corresponding non-invertible CA when $\alpha_{[H_i]}(G; A) = 1$.

Theorem 4. *Let G be a finite abelian group and A a finite set of size $q \geq 2$. Let $H_1, H_2, \ldots, H_r$ be the list of different subgroups of G with $H_1 = G$. For each $1 \leq i \leq r$, let $\alpha_i := \alpha_{[H_i]}(G; A)$. Then,*

$$\mathrm{Rank}(\mathrm{CA}(G; A) : \mathrm{ICA}(G; A)) = |\mathcal{E}_G| - \sum_{i=2}^{r} \delta(\alpha_i, 1),$$

where $\delta : \mathbb{N}^2 \to \{0, 1\}$ is Kronecker's delta function.

Proof. For all $1 \leq i \leq r$, let $B_i := B_{[H_i]}$. Fix orbits $x_i G \subseteq B_i$, so $H_i = G_{x_i}$. Assume that the list of subgroups of G is ordered such that

$$|x_1 G| \leq \cdots \leq |x_r G|, \text{ or, equivalently, } |G_{x_1}| \geq \cdots \geq |G_{x_r}|.$$

For every $\alpha_i \geq 2$, fix orbits $y_i G \subseteq B_i$ such that $x_i G \neq y_i G$. We claim that $\mathrm{CA}(G, A) = M := \langle \mathrm{ICA}(G; A) \cup U \rangle$, where

$$U := \left\{ \tau_{x_i, x_j} : [G_{x_i}] < [G_{x_j}] \right\} \cup \left\{ \tau_{x_i, y_i} : \alpha_i \geq 2 \right\},$$

and $\tau_{x_i, x_j}, \tau_{x_i, y_i}$ are the idempotents defined in Corollary 3. For any $\tau \in \mathrm{CA}(G; A)$, consider $\tau_i \in \mathrm{CA}(G; A)$, $1 \leq i \leq r$, defined by

$$(x)\tau_i = \begin{cases} (x)\tau & \text{if } x \in B_i \\ x & \text{otheriwse.} \end{cases}$$

By Lemmas 3 and 6, $(B_i)\tau \subseteq \bigcup_{j \leq i} B_j$ for all i. Hence, we have the decomposition

$$\tau = \tau_1 \tau_2 \ldots \tau_r.$$

We shall prove that $\tau_i \in M$ for all $1 \leq i \leq r$. For each i, decompose $B_i = B_i' \cup B_i''$, where

$$B_i' := \bigcup \{ P \in \mathcal{O}(G; A) : P \subseteq B_i \text{ and } (P)\tau_i \subseteq B_j \text{ for some } j < i \},$$
$$B_i'' := \bigcup \{ P \in \mathcal{O}(G; A) : P \subseteq B_i \text{ and } (P)\tau_i \subseteq B_i \}.$$

If τ_i' and τ_i'' are the transformations that act as τ_i on B_i' and B_i'', respectively, and fix everything else, then $\tau_i = \tau_i' \tau_i''$. We shall prove that $\tau_i' \in M$ and $\tau_i'' \in M$.

1. We show that $\tau_i' \in M$. For any orbit $P \in B_i'$, the orbit $Q := (P)\tau_i'$ is contained in B_j for some $j < i$. By Theorem 2, there exist

$$\phi \in \left((G/G_{x_i}) \wr \mathrm{Sym}_{\alpha_i} \right) \times \left((G/G_{x_j}) \wr \mathrm{Sym}_{\alpha_j} \right) \leq \mathrm{ICA}(G; A)$$

such that ϕ acts as the double transposition $(x_iG, P)(x_jG, Q)$. Since G/G_{x_i} and G/G_{x_j} are transitive on x_iG and x_jG, respectively, we may take ϕ such that $(x_i)\phi\tau_i' = (x_j)\phi$. Then,

$$(z)\tau_i' = (z)\phi\tau_{x_i,x_j}\phi, \ \forall z \in P.$$

As τ_i' may be decomposed as a product of transformations that only move one orbit in B_i', this shows that $\tau_i' \in M$.

2. We show that $\tau_i'' \in M$. In this case, $\tau_i'' \in \mathrm{Tran}(B_i)$. In fact, as τ_i'' preserves the partition of B_i into G-orbits, Lemma 5 implies that $\tau_i'' \in (G/G_{x_i}) \wr \mathrm{Tran}_{\alpha_i}$. If $\alpha_i \geq 2$, the semigroup Tran_{α_i} is generated by $\mathrm{Sym}_{\alpha_i} \leq \mathrm{ICA}(G, A)$ together with the idempotent τ_{x_i,y_i}. Hence, $\tau_i'' \in M$.

Therefore, we have established that $\mathrm{CA}(G; A) = \langle \mathrm{ICA}(G; A) \cup U \rangle$.
Suppose now that there exists $V \subseteq \mathrm{CA}(G; A)$ such that $|V| < |U|$ and

$$\langle \mathrm{ICA}(G; A) \cup V \rangle = \mathrm{CA}(G; A).$$

The elements of U have all different kernels, so, for some $\tau \in U$, we must have

$$V \cap \langle \mathrm{ICA}(G; A), \tau \rangle = \emptyset.$$

(Otherwise, V has at least $|U|$ transformations with different kernels, which is impossible since $|V| < |U|$). If $\tau = \tau_{x_i,y_i}$, for some i with $\alpha_i \geq 2$, this implies that there is no $\xi \in V$ with

$$\ker(\xi) = \{\{a,b\}, \{c\} : a \in x_iG, \ b \in y_iG, \ c \in A^G \setminus (x_iG \cup y_iG)\}.$$

Hence, there is no $\xi \in \langle \mathrm{ICA}(G; A) \cup V \rangle = \mathrm{CA}(G; A)$ with kernel of this form, which is a contradiction because τ_{x_i,y_i} itself has kernel of this form. We obtain a similar contradiction if $\tau = \tau_{x_i,x_j}$ with $[G_{x_i}] < [G_{x_j}]$. $\qquad\square$

Corollary 4. *Let G be a finite abelian group with $\mathrm{Rank}(G) = m$ and A a finite set of size $q \geq 2$. With the notation of Theorem 4,*

$$\mathrm{Rank}(\mathrm{CA}(G; A)) \leq \sum_{i=2}^{r} m\alpha_i + 2r + |\mathcal{E}_G| - \delta(q,2) - \sum_{i=2}^{r}(3\delta(\alpha_i,1) + \delta(\alpha_i,2))$$

$$\leq \sum_{i=2}^{r} m\alpha_i + 2r + r^2.$$

Proof. Using the fact $\mathrm{Rank}((G/H_i) \wr \mathrm{Sym}_{\alpha_i}) \leq m\alpha_i + 2 - 2\delta(\alpha_i, 1) - \delta(\alpha_i, 2)$ and $\mathrm{Rank}((G/H_1) \wr \mathrm{Sym}_q) = 2 - \delta(q, 2)$, the result follows by Theorem 4, Corollary 2 and Lemma 4. $\qquad\square$

The bound of Corollary 4 may become tighter if we actually know $\mathrm{Rank}(G/H_i)$, for all $H_i \leq G$, as in Example 2.

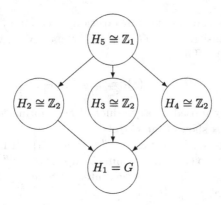

Fig. 1. Hasse diagram of subgroup lattice of $G = \mathbb{Z}_2 \times \mathbb{Z}_2$.

Example 3. Let $G = \mathbb{Z}_2 \times \mathbb{Z}_2$ be the Klein-four group and $A = \{0, 1\}$. With the notation of Example 1, Fig. 1 illustrates the Hasse diagram of the subgroup lattice of G (i.e. the actual lattice of subgroups is the transitive and reflexive closure of this graph). Hence, by Theorem 4 and Example 2,

$$\text{Rank}(\text{CA}(G; A) : \text{ICA}(G; A)) = |\mathcal{E}_G| - 3 = 12 - 3 = 9,$$
$$\text{Rank}(\text{CA}(G; A)) \leq 9 + 9 = 18, \text{ as } \text{Rank}(\text{ICA}(G; A)) \leq 9.$$

Because of Theorem 4, it is particularly relevant to determine in which situations $\alpha_{[H]}(G; A) = 1$. We finish this paper with some partial results in this direction that hold for arbitrary finite groups.

Denote by $[G : H]$ the index of $H \leq G$ (i.e. the number of cosets of H in G).

Lemma 7. *Let G be a finite group and A a finite set of size $q \geq 2$. Assume there is $H \leq G$ with $[G : H] = 2$. Then, $\alpha_{[H]}(G; A) = 1$ if and only if $q = 2$.*

Proof. As $H \leq G$ has index 2, it is normal. Fix $s \in G \setminus H$. Define $x \in A^G$ by

$$(g)x = \begin{cases} 0 & \text{if } g \in H \\ 1 & \text{if } g \in sH = Hs. \end{cases}$$

Clearly $G_x = H$ and $x \in B_{[H]}$.

Suppose first that $A = \{0, 1\}$. Let $y \in B_{[H]}$. As H is normal, $[H] = \{H\}$, so $G_y = H$. For any $h \in H$,

$$(h)y = (e)y \cdot h^{-1} = (e)y \text{ and } (sh)y = (s)y \cdot h^{-1} = (s)y,$$

so y is constant on the cosets H and $sH = Hs$. Therefore, either $y = x$, or

$$(g)y = \begin{cases} 1 & \text{if } g \in H \\ 0 & \text{if } g \in sH = Hs. \end{cases}$$

In the latter case, $y \cdot s = x$ and $y \in xG$. This shows that there is a unique G-orbit contained in $B_{[H]}$, so $\alpha_{[H]}(G; A) = 1$.

If $|A| \geq 3$, we may use a similar argument as above, except that now $y \in B_{[H]}$ may satisfy $(g)y \in A \setminus \{0, 1\}$ for all $g \in H$, so $y \notin xG$ and $\alpha_{[H]}(G; A) \geq 2$.   □

Lemma 8. *Let G be a finite group and A a finite set of size $q \geq 2$. Suppose there is $H \leq G$ such that $\alpha_{[H]}(G; A) = 1$. Then, $q \mid [G : H] = \frac{|G|}{|H|}$.*

Proof. Let $x \in B_{[H]}$ be such that $G_x = H$. As $\alpha_{[H]}(G; A) = 1$, $B_{[H]} = xG$. First we show that $x : G \to A$ is surjective. If $(G)x \subset A$, let $a \in (G)x$ and $b \in A \setminus (G)x$. Define $y \in A^G$ by

$$(g)y := \begin{cases} b & \text{if } (g)x = a \\ (g)x & \text{otherwise.} \end{cases}$$

Then $y \in B_{[H]}$, as $G_y = G_x$, but $y \notin xG$, which is a contradiction. For $a \in A$, let $(a)x^{-1} := \{g \in G : (g)x = a\}$. Now we show that, for any $a, b \in A$,

$$|(a)x^{-1}| = |(b)x^{-1}|.$$

Suppose that $|(a)x^{-1}| < |(b)x^{-1}|$. Define $z \in A^G$ by

$$(g)z := \begin{cases} b & \text{if } (g)x = a \\ a & \text{if } (g)x = b \\ (g)x & \text{otherwise.} \end{cases}$$

Again, $z \in B_{[H]}$, as $G_z = G_x$, but $z \notin xG$, which is a contradiction. As x is constant on the left cosets of H in G, for each $a \in A$, $(a)x^{-1}$ is a union of left cosets. All cosets have the same size, so $(a)x^{-1}$ and $(b)x^{-1}$ contain the same number of them, for any $a, b \in A$. Therefore, $q \mid [G : H]$.   □

Corollary 5. *Let G be a finite abelian group and A a finite set of size $q \geq 2$ such that $q \nmid |G|$. With the notation of Theorem 4,*

$$\text{Rank}(\text{CA}(G; A) : \text{ICA}(G; A)) = |\mathcal{E}_G|.$$

Acknowledgments. This work was supported by the EPSRC grant EP/K033956/1.

References

1. Araújo, J., Bentz, W., Mitchell, J.D., Schneider, C.: The rank of the semigroup of transformations stabilising a partition of a finite set. Mat. Proc. Camb. Phil. Soc. **159**, 339–353 (2015)
2. Araújo, J., Schneider, C.: The rank of the endomorphism monoid of a uniform partition. Semigroup Forum **78**, 498–510 (2009)
3. Castillo-Ramirez, A., Gadouleau, M.: Ranks of finite semigroups of one-dimensional cellular automata. Semigroup Forum (First online), pp. 1–16 (2016). doi:10.1007/s00233-016-9783-z

4. Ceccherini-Silberstein, T., Coornaert, M.: Cellular Automata and Groups. Springer Monographs in Mathematics. Springer, Heidelberg (2010)
5. Dixon, J.D., Mortimer, B.: Permutation Groups. Graduate Texts in Mathematics, vol. 163. Springer, New York (1996)
6. Ganyushkin, O., Mazorchuk, V.: Classical Finite Transformation Semigroups: An Introduction. Algebra and Applications, vol. 9. Springer, London (2009)
7. Gomes, G.M.S., Howie, J.M.: On the ranks of certain finite semigroups of transformations. Math. Proc. Camb. Phil. Soc. **101**, 395–403 (1987)
8. Gray, R.D.: The minimal number of generators of a finite semigroup. Semigroup Forum **89**, 135–154 (2014)
9. Hartman, Y.: Large semigroups of cellular automata. Ergodic Theory Dyn. Syst. **32**, 1991–2010 (2012)
10. Howie, J.M., McFadden, R.B.: Idempotent rank in finite full transformation semigroups. Proc. Royal Soc. Edinburgh **114A**, 161–167 (1990)
11. Kari, J.: Theory of cellular automata: a survey. Theoret. Comput. Sci. **334**, 3–33 (2005)
12. Salo, V.: Groups and monoids of cellular automata. In: Kari, J. (ed.) AUTOMATA 2015. LNCS, vol. 9099, pp. 17–45. Springer, Heidelberg (2015)

Sum of Exit Times in Series of Metastable States in Probabilistic Cellular Automata

E.N.M. Cirillo[1], F.R. Nardi[2,3], and C. Spitoni[4(✉)]

[1] Dipartimento di Scienze di Base e Applicate per l'Ingegneria,
Sapienza Università di Roma, via A. Scarpa 16, 00161 Roma, Italy
emilio.cirillo@uniroma1.it
[2] Department of Mathematics and Computer Science, Eindhoven University
of Technology, P.O. Box 513, 5600 MB Eindhoven, The Netherlands
F.R.Nardi@tue.nl
[3] Eurandom, Eindhoven University, P.O. Box 513,
5600 MB Eindhoven, The Netherlands
[4] Department of Mathematics, Utrecht University, Budapestlaan 6,
3584 CD Utrecht, The Netherlands
C.Spitoni@uu.nl

Abstract. Reversible Probabilistic Cellular Automata are a special class of automata whose stationary behavior is described by Gibbs-like measures. For those models the dynamics can be trapped for a very long time in states which are very different from the ones typical of stationarity. This phenomenon can be recast in the framework of metastability theory which is typical of Statistical Mechanics. In this paper we consider a model presenting two not degenerate in energy metastable states which form a series, in the sense that, when the dynamics is started at one of them, before reaching stationarity, the system must necessarily visit the second one. We discuss a rule for combining the exit times from each of the metastable states.

1 Introduction

Cellular Automata (CA) are discrete-time dynamical systems on a spatially extended discrete space, see, e.g., [8] and references therein. Probabilistic Cellular Automata (PCA) are CA straightforward generalizations where the updating rule is stochastic (see [10,13,17]). Strong relations exist between PCA and the general equilibrium statistical mechanics framework [7,10]. Traditionally, the interplay between disordered global states and ordered phases has been addressed, but, more recently, it has been remarked that even from the non-equilibrium point of view analogies between statistical mechanics systems and PCA deserve attention [2].

In this paper we shall consider a particular class of PCA, called *reversible* PCA. Here the word reversible is used in the sense that the detailed balance

© IFIP International Federation for Information Processing 2016
Published by Springer International Publishing Switzerland 2016. All Rights Reserved
M. Cook and T. Neary (Eds.): AUTOMATA 2016, LNCS 9664, pp. 105–119, 2016.
DOI: 10.1007/978-3-319-39300-1_9

condition is satisfied with respect to a suitable Gibbs-like measure (see the precise definition given just below equation (2.3)) defined via a translation invariant multi-body potential. Such a measure depends on a parameter which plays a role similar to that played by the temperature in the context of statistical mechanics systems and which, for such a reason, will be called temperature. In particular, for small values of such a temperature, the dynamics of the PCA tends to be frozen in the local minima of the Hamiltonian associated to the Gibbs-like measure. Moreover, in suitable low temperature regimes (see [11]) the transition probabilities of the PCA become very small and the effective change of a cell's state becomes rare, so that the PCA dynamics becomes almost a sequential one.

It is natural to pose, even for reversible PCA's, the question of metastability which arose overbearingly in the history of thermodynamics and statistical mechanics since the pioneering works due to van der Waals.

Metastable states are observed when a physical system is close to a first order phase transition. Well-known examples are super-saturated vapor states and magnetic hysteresis [16]. Not completely rigorous approaches based on equilibrium states have been developed in different fashions. However, a fully mathematically rigorous theory, has been obtained by approaching the problem from a dynamical point of view. For statistical mechanics systems on a lattice a dynamics is introduced (a Markov process having the Gibbs measure as stationary measure) and metastable states are interpreted as those states of the system such that the corresponding time needed to relax to equilibrium is the longest one on an exponential scale controlled by the inverse of the temperature. The purely dynamical point of view revealed itself extremely powerful and led to a very elegant definition and characterization of the metastable states. The most important results in this respect have been summed up in [16].

The dynamical description of metastable states suits perfectly for their generalization to PCA [2–5]. Metastable states have been investigated for PCA's in the framework of the so called *pathwise approach* [12,15,16]. It has been shown how it is possible to characterize the exit time from the metastable states up to an exponential precision and the typical trajectory followed by the system during its transition from the metastable to the stable state. Moreover, it has also been shown how to apply the so called *potential theoretic approach* [1] to compute sharp estimates for the exit time [14] of a specific PCA.

More precisely, the exit time from the metastable state is essentially in the form $K \exp\{\Gamma/T\}$ where T is the temperature, Γ is the energy cost of the best (in terms of energy) paths connecting the metastable state to the stable one, and K is a number which is inversely connected to the number of possible best paths that the system can follow to perform its transition from the metastable state to the stable one. Up to now, in the framework of PCA models, the constant K has been computed only in cases in which the metastable state is unique. The aim of this work is to consider a PCA model in which two metastable states are present. Similar results in the framework of the Blume-Capel model with Metropolis dynamics have been proved in [6,9].

We shall consider the PCA studied in [2] which is characterized by the presence of two metastable states. Moreover, starting from one of them, the system, in order to perform its transition to the stable state, must necessarily visit the second metastable state. The problem we pose and solve in this paper is that of studying how the exit times from the two metastable states have to be combined to find the constant K characterizing the transition from the first metastable state to the stable one. We prove that K is the sum of the two constants associated with the exit times from the two metastable states.

The paper is organized as follows, in Sect. 2 we define the class of models considered, in Sect. 3 we state the main result and recall the main mathematical tools used in its proof, and in Sect. 4 we sketch the proof of the main theorem of the paper.

2   The Model

In this section we introduce the basic notation and we define the model of reversible PCA which will be studied in the sequel. Consider the two-dimensional torus $\Lambda = \{0, \ldots, L-1\}^2$, with L even[1], endowed with the Euclidean metric. Associate a variable $\sigma(x) = \pm 1$ with each site $x \in \Lambda$ and let $\mathcal{S} = \{-1, +1\}^\Lambda$ be the configuration space. Let $\beta > 0$ and $h \in (0, 1)$. Consider the Markov chain σ_n, with $n = 0, 1, \ldots$, on $\mathcal{S}$ with transition matrix:

$$p(\sigma, \eta) = \prod_{x \in \Lambda} p_{x,\sigma}(\eta(x)) \quad \forall \sigma, \eta \in \mathcal{S} \tag{2.1}$$

where, for $x \in \Lambda$ and $\sigma \in \mathcal{S}$, $p_{x,\sigma}(\cdot)$ is the probability measure on $\{-1, +1\}$ defined as

$$p_{x,\sigma}(s) = \frac{1}{1 + \exp\{-2\beta s(S_\sigma(x) + h)\}} = \frac{1}{2}[1 + s \tanh \beta (S_\sigma(x) + h)] \tag{2.2}$$

with $s \in \{-1, +1\}$ and $S_\sigma(x) = \sum_{y \in \Lambda} K(x - y)\sigma(y)$ where $K(x - y) = 1$ if $|x - y| = 1$, and $K(x - y) = 0$ otherwise. The probability $p_{x,\sigma}(s)$ for the spin $\sigma(x)$ to be equal to s depends only on the values of the spins of σ on the diamond $V(x)$ centered at x, as shown in Fig. 1 (i.e., the von Neumann neighborhood without the center).

At each step of the dynamics all the spins of the system are updated simultaneously according to the probability distribution (2.2). This means the value of the spin tends to align with the local field $S_\sigma(x) + h$: $S_\sigma(x)$ mimics a ferromagnetic interaction effect among spins, whereas h is an external *magnetic field*. Such a field, as said before, is chosen smaller than one otherwise its effect would be so strong to destroy the metastable behavior. When β is large the tendency to align with the local field is perfect, while for β small also spin updates against the local filed can be observed with a not too small probability. Thus β can be interpreted as the inverse of the *temperature*.

---

[1] The side length of the lattice is chosen even so that it will possible to consider configurations in which the plus and the minus spins for a checkerboard and fulfill the periodic boundary conditions.

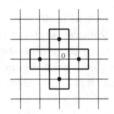

Fig. 1. Diamond $V(0)$ for the nearest neighbor model.

This kernel K choice leads to the *nearest neighbor PCA* model studied in [2]. The Markov chain σ_n defined in (2.1) updates all the spins simultaneously and independently at any time and it satisfies the *detailed balance* property $p(\sigma, \eta) \, e^{-\beta H(\sigma)} = p(\eta, \sigma) \, e^{-\beta H(\eta)}$ with

$$H(\sigma) = -h \sum_{x \in \Lambda} \sigma(x) - \frac{1}{\beta} \sum_{x \in \Lambda} \log \cosh \left[\beta \left(S_\sigma(x) + h \right) \right]. \tag{2.3}$$

This is also expressed by saying that the dynamics is *reversible* with respect to the Gibbs measure $\mu(\sigma) = \exp\{-\beta H(\sigma)\}/Z$ with $Z = \sum_{\eta \in \mathcal{S}} \exp\{-\beta H(\eta)\}$. This property implies that μ is stationary, i.e., $\sum_{\sigma \in \mathcal{S}} \mu(\sigma) p(\sigma, \eta) = \mu(\eta)$.

It is important to remark that, although the dynamics is reversible, the probability $p(\sigma, \eta)$ cannot be expressed in terms of $H(\sigma) - H(\eta)$, as it usually happens for the serial Glauber dynamics, typical of Statistical Mechanics. Thus, given $\sigma, \eta \in \mathcal{S}$, we define the *energy cost*

$$\Delta(\sigma, \eta) = - \lim_{\beta \to \infty} \frac{\log p(\sigma, \eta)}{\beta} = \sum_{\substack{x \in \Lambda: \\ \eta(x)[S_\sigma(x) + h] < 0}} 2|S_\sigma(x) + h| \tag{2.4}$$

Note that $\Delta(\sigma, \eta) \geq 0$ and $\Delta(\sigma, \eta)$ is not necessarily equal to $\Delta(\eta, \sigma)$; it can be proven, see [3, Sect. 2.6], that

$$e^{-\beta \Delta(\sigma,\eta) - \beta \gamma(\beta)} \leq p(\sigma, \eta) \leq e^{-\beta \Delta(\sigma,\eta) + \beta \gamma(\beta)} \tag{2.5}$$

with $\gamma(\beta) \to 0$ in the zero temperature limit $\beta \to \infty$. Hence, Δ can be interpreted as the cost of the transition from σ to η and plays the role that, in the context of Glauber dynamics, is played by the difference of energy. In this context the ground states are those configurations on which the Gibbs measure μ concentrates when $\beta \to \infty$; hence, they can be defined as the minima of the *energy*:

$$E(\sigma) = \lim_{\beta \to \infty} H(\sigma) = -h \sum_{x \in \Lambda} \sigma(x) - \sum_{x \in \Lambda} |S_\sigma(x) + h| \tag{2.6}$$

For $h > 0$ the configuration $+\underline{1}$, with $+\underline{1}(x) = +1$ for $x \in \Lambda$, is the unique ground state, indeed each site contributes to the energy with $-h - (4 + h)$. For $h = 0$, the configuration $-\underline{1}$, with $-\underline{1}(x) = -1$ for $x \in \Lambda$, is a ground state as well, as all the other configurations such that all the sites contribute to the

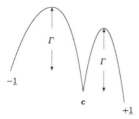

Fig. 2. Schematic description of the energy landscape for a series of metastable states. Note that the ground state is $+\underline{1}$ and $E(-\underline{1}) > E(\mathbf{c}) > E(+\underline{1})$.

sum (2.6) with 4. Hence, the checkerboard configurations $\mathbf{c}^{\mathrm{e}}, \mathbf{c}^{\mathrm{o}} \in \mathcal{S}$ such that $\mathbf{c}^{\mathrm{e}}(x) = (-1)^{x_1+x_2}$ and $\mathbf{c}^{\mathrm{o}}(x) = (-1)^{x_1+x_2+1}$ for $x = (x_1, x_2) \in \Lambda$ are ground states, as well. Notice that $\mathbf{c}^{\mathrm{e}}$ and $\mathbf{c}^{\mathrm{o}}$ are checkerboard-like states with the pluses on the even and odd sub-lattices, respectively; we set $\mathbf{c} = \{\mathbf{c}^{\mathrm{e}}, \mathbf{c}^{\mathrm{o}}\}$. Since the side length L of the torus Λ is even, then $E(\mathbf{c}^{\mathrm{e}}) = E(\mathbf{c}^{\mathrm{o}}) = E(\mathbf{c})$ (we stress the abuse of notation $E(\mathbf{c})$). Under periodic boundary conditions, we get for the energies: $E(+\underline{1}) = -L^2(4 + 2h)$, $E(-\underline{1}) = -L^2(4 - 2h)$, and $E(\mathbf{c}) = -4L^2$. Therefore,

$$E(-\underline{1}) > E(\mathbf{c}) > E(+\underline{1}) \tag{2.7}$$

for $0 < h \leq 1$. Moreover, using the analysis in [2] we can derive Fig. 2, with the series of the two local minima $-\underline{1}, \mathbf{c}$.

We conclude this section by listing some relevant definitions. Given $\sigma \in \mathcal{S}$ we consider the chain with initial configuration $\sigma_0 = \sigma$, we denote with $\mathbb{P}_\sigma$ the probability measure on the space of trajectories, by $\mathbb{E}_\sigma$ the corresponding expectation value, and by

$$\tau_A^\sigma := \inf\{t > 0 : \sigma_t \in A\} \tag{2.8}$$

the *first hitting time on* $A \subset \mathcal{S}$; we shall drop the initial configuration from the notation (2.8) whenever it is equal to $-\underline{1}$, we shall write τ_A for $\tau_A^{-\underline{1}}$, namely. Moreover, a finite sequence of configurations $\omega = \{\omega_1, \ldots, \omega_n\}$ is called the *path* with starting configuration ω_1 and ending configuration ω_n; we let $|\omega| := n$.

Given a path ω we define the *height* along ω as:

$$\Phi_\omega := H(\omega_1) \text{ if } |\omega| = 1 \quad \text{and} \quad \Phi_\omega := \max_{i=1,\ldots,|\omega|-1} H(\omega_i, \omega_{i+1}) \quad \text{otherwise} \tag{2.9}$$

where $H(\omega_i, \omega_{i+1})$ is the *communication height* between the configurations ω_i and ω_{i+1}, defined as follows:

$$H(\omega_i, \omega_{i+1}) := H(\omega_i) - \frac{1}{\beta} \log(p(\omega_i, \omega_{i+1})) \tag{2.10}$$

Given two configurations $\sigma, \eta \in \mathcal{S}$, we denote by $\Theta(\sigma, \eta)$ the set of all the paths ω starting from σ and ending in η. The minimax between σ and η is defined as

$$\Phi(\sigma, \eta) := \min_{\omega \in \Theta(\sigma, \eta)} \Phi_\omega \tag{2.11}$$

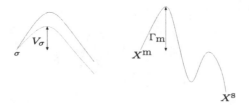

Fig. 3. Definition of metastable states.

3   Metastable States and Main Results

We want now to define the notion of metastable states. See Fig. 3 for a graphic description of the quantities we are going to define. For any $\sigma \in \mathcal{S}$, we let $\mathcal{I}_\sigma \subset \mathcal{S}$ be the set of configurations with energy strictly below $H(\sigma)$ and $V_\sigma = \Phi(\sigma, \mathcal{I}_\sigma) - H(\sigma)$ be the *stability level of* σ, that is the energy barrier that, starting from σ, must be overcome to reach the set of configurations with energy smaller than $H(\sigma)$. We denote by X^s the set of global minima of the energy, i.e., the collection of the ground states, and suppose that the *communication energy* $\Gamma = \max_{\sigma \in \mathcal{S} \setminus X^\text{s}} V_\sigma$ is strictly positive. Finally, we define the set of *metastable states* $X^\text{m} = \{\eta \in \mathcal{S} : V_\eta = \Gamma\}$. The set X^m deserves its name, since in a rather general framework it is possible to prove (see, e.g., [12, Theorem 4.9]) the following: pick $\sigma \in X^\text{m}$, consider the chain σ_n started at $\sigma_0 = \sigma$, then the *first hitting time* τ_{X^s} to the ground states is a random variable with mean exponentially large in β, that is

$$\lim_{\beta \to \infty} \frac{1}{\beta} \log \mathbb{E}_\sigma[\tau_{X^\text{s}}] = \Gamma \tag{3.12}$$

In the considered regime, finite volume and temperature tending to zero, the description of metastability is then reduced to the computation of X^s, Γ, and X^m.

We pose now the problem of metastability and state the related theorem on the sharp estimates for the exit time. Consider the model (2.1) with $0 < h < 1$ and suppose that the system is prepared in the state $\sigma_0 = -\underline{1}$, and we estimate the first time at which the system reaches $+\underline{1}$. As showed in [2], the system visits with probability tending to one in the $\beta \to \infty$ limit the checkerboards **c**, and the typical time to jump from $-\underline{1}$ to **c** is the same as the time needed to jump from **c** to $+\underline{1}$. Hence, the aim of this paper is to prove an addition formula for the expected exit times from $-\underline{1}$ to $+\underline{1}$. The metastable states $-\underline{1}$ and **c** form indeed a *series*: the system started at $-\underline{1}$ must necessarily pass through **c** before relaxing to the stable state $+\underline{1}$.

In order to state the main theorem, we have to introduce the following *activation energy* Γ_m, which corresponds to the energy of the critical configuration triggering the nucleation:

$$\Gamma_\text{m} := -2h\lambda^2 + 2(4+h)\lambda - 2h \tag{3.13}$$

where λ is the *critical length* defined as $\lambda := \lfloor 2/h \rfloor + 1$. In other words, in [2] it is proven that with probability tending to one in the limit $\beta \to \infty$, before reaching the checkerboards $\mathbf{c}$, the system necessarily visits a particular set of configurations, called *critical droplets*, which are all the configurations (equivalent up to translations, rotations, and protuberance possible shifts) with a single checkerboard droplet in the sea of minuses with the following shape: a rectangle of side lengths λ and $\lambda - 1$ with a unit square attached to one of the two longest sides (the protuberance) and with the spin in the protuberance being plus. This way of escaping from the metastable via the formation of a single droplet is called *nucleation*.

In order to study the exit times from the from one of the two metastable states towards the stable configuration, we use a decomposition valid for a general Markov chains derived in [6] and that we recall for the sake of completeness.

Lemma 3.1. *Consider a finite state space X and a family of irreducible and aperiodic Markov chain. Given three states $y, w, z \in X$ pairwise mutually different, we have that the following holds*

$$\mathbb{E}_y[\tau_z] = \mathbb{E}_y[\tau_w \mathbf{1}_{\{\tau_w < \tau_z\}}] + \mathbb{E}_w[\tau_z]\mathbb{P}_y(\tau_w < \tau_z) + \mathbb{E}_y[\tau_z \mathbf{1}_{\{\tau_w \geq \tau_z\}}] \qquad (3.14)$$

where $\mathbb{E}_x[\cdot]$ denotes the average along the trajectories of the Markov chain started at x, and τ_y is the first hitting time to y for the chain started at x. In the expressions above $\mathbf{1}_{\{\cdot\}}$ is the characteristic function which is equal to one if the event $\{\cdot\}$ is realized and zero otherwise.

Theorem 3.1. *Consider the PCA (2.1), for $h > 0$ small enough, and β large enough, we have:*

$$\mathbb{E}_{-\underline{1}}(\tau_{+\underline{1}}) = \left(\frac{1}{k_1} + \frac{1}{k_2} \right) e^{\beta \Gamma_m}[1 + o(1)], \qquad (3.15)$$

where $o(1)$ denotes a function going to zero when $\beta \to \infty$ and

$$k_1 = k_2 = K = 8\lambda|\Lambda|$$

The term $e^{\beta \Gamma_m}/k_1$ in (3.15) represents the contribution of the mean hitting time $\mathbb{E}_{-\underline{1}}[\tau_{\mathbf{c}} \mathbf{1}_{\{\tau_{\mathbf{c}} < \tau_{+\underline{1}}\}}]$ to the relation (3.14) and $e^{\beta \Gamma_m}/k_2$ the contribution of $\mathbb{E}_{\mathbf{c}} \tau_{+\underline{1}}$. The pre-factors k_1 and k_2 give the precise estimate of the mean nucleation time of the stable phase, beyond the asymptotic exponential regime and represent entropic factors, counting the cardinality of the critical droplets which trigger the nucleation. At the level of logarithmic equivalence, namely, by renouncing to get sharp estimate, this result can be proven by the methods in [12]. More precisely, one gets that $(1/\beta) \log \mathbb{E}_{-\underline{1}}[\tau_{+\underline{1}}]$ tends to Γ_m in the large β limit.

3.1   Potential Theoretic Approach and Capacities

Since our results on the precise asymptotic of the mean nucleation time of the stable phase are strictly related to the potential theoretic approach to metastability (see [1]), we recall some definitions and notions. We define the *Dirichlet form* associated to the reversible Markov chain, with transition probabilities $p(\sigma, \eta)$ and equilibrium measure μ, as the functional:

$$\mathcal{E}(h) = \frac{1}{2} \sum_{\sigma, \eta \in S} \mu(\sigma) p(\sigma, \eta)[f(\sigma) - f(\eta)]^2, \tag{3.16}$$

where $f : S \to [0, 1]$ is a generic function. The form (3.16) can be rewritten in terms of the communication heights $H(\sigma, \eta)$ and of the partition function Z:

$$\mathcal{E}(h) = \frac{1}{2} \sum_{\sigma, \eta \in S} \frac{1}{Z} e^{-\beta H(\sigma, \eta)}[f(\sigma) - f(\eta)]^2. \tag{3.17}$$

Given two non-empty disjoint sets $\mathcal{A}, \mathcal{B}$ the *capacity* of the pair $\mathcal{A}, \mathcal{B}$ is defined by

$$\text{cap}_\beta(\mathcal{A}, \mathcal{B}) := \min_{\substack{f:S \to [0,1] \\ f|_\mathcal{A}=1, f|_\mathcal{B}=0}} \mathcal{E}(f) \tag{3.18}$$

and from this definition it follows that the capacity is a *symmetric* function of the sets $\mathcal{A}$ and $\mathcal{B}$.

The right hand side of (3.18) has a unique minimizer $f^\star_{\mathcal{A},\mathcal{B}}$ called *equilibrium potential* of the pair $\mathcal{A}, \mathcal{B}$ given by

$$f^\star_{\mathcal{A},\mathcal{B}}(\eta) = \mathbb{P}_\eta(\tau_\mathcal{A} < \tau_\mathcal{B}), \tag{3.19}$$

for any $\eta \notin \mathcal{A} \cup \mathcal{B}$. The strength of this variational representation comes from the *monotonicity* of the Dirichlet form in the variable $p(\sigma, \eta)$. In fact, the Dirichlet form $\mathcal{E}(f)$ is a monotone increasing function of the transition probabilities $p(x, y)$ for $x \neq y$, while it is independent on the value $p(x, x)$. In fact, the following theorem holds:

Theorem 3.2. *Assume that $\mathcal{E}$ and $\tilde{\mathcal{E}}$ are Dirichlet forms associated to two Markov chains P and $\tilde{P}$ with state space S and reversible with respect to the measure μ. Assume that the transition probabilities p and $\tilde{p}$ are given, for $x \neq y$, by*

$$p(x, y) = g(x, y)/\mu(x) \quad and \quad \tilde{p}(x, y) = \tilde{g}(x, y)/\mu(x)$$

where $g(x, y) = g(y, x)$ and $\tilde{g}(x, y) = \tilde{g}(y, x)$, and, for all $x \neq y$, $\tilde{g}(x, y) \leq g(x, y)$. Then, for any disjoint sets $\mathcal{A}, \mathcal{D} \subset S$ we have:

$$\text{cap}_\beta(\mathcal{A}, \mathcal{D}) \geq \widetilde{\text{cap}}_\beta(\mathcal{A}, \mathcal{D}) \tag{3.20}$$

We will use Theorem 3.2 by simply setting some of the transition probabilities $p(x, y)$ equal to zero. Indeed if enough of these are zero, we obtain a chain where everything can be computed easily. In order to get a good lower bound, the trick will be to guess which transitions can be switched off without altering the capacities too much, and still to simplify enough to be able to compute it.

3.2   Series of Metastable States for PCA Without Self-interaction

In this section we state the model-dependent results for the class of PCA considered, which, by the general theory contained in [6], imply Theorem 3.1.

Lemma 3.2. *The configurations $-\underline{1}$, $\mathbf{c}$, and $+\underline{1}$ are such that $X^{\mathrm{s}} = \{+\underline{1}\}$, $X^{\mathrm{m}} = \{-\underline{1}, \mathbf{c}\}$, $E(-\underline{1}) > E(\mathbf{c})$, and $\Gamma = \Gamma_{\mathrm{m}}$.*

Our model presents the series structure depicted in Fig. 2: when the system is started at $-\underline{1}$ with high probability it will visit $\mathbf{c}$ before $+\underline{1}$. In fact, the following lemma holds:

Lemma 3.3. *There exists $\lambda > 0$ and $\beta_0 > 0$ such that for any $\beta > \beta_0$*

$$\mathbb{P}_{-\underline{1}}(\tau_{+\underline{1}} < \tau_{\mathbf{c}}) \leq e^{-\beta\lambda} \tag{3.21}$$

We use Lemmas 3.1 and 3.3 and, in order to drop the last addendum of (3.1), we need an exponential control of the tail of the distribution of the suitably rescaled random variable τ_{x_0}.

Lemma 3.4. *For any $\delta > 0$ there exists $\beta_0 > 0$ such that*

$$\sum_{t=0}^{\infty} t\, \mathbb{P}_{-\underline{1}}(\tau_{+\underline{1}} > te^{\beta\Gamma_{\mathrm{m}}+\beta\delta}) \leq 1/3 \tag{3.22}$$

for any $\beta > \beta_0$.

By Lemmas 3.1, 3.3 and 3.4 we have that:

$$\mathbb{E}_{-\underline{1}}[\tau_{+\underline{1}}] = \left(\mathbb{E}_{-\underline{1}}[\tau_{\mathbf{c}}\mathbf{1}_{\{\tau_{\mathbf{c}}<\tau_{+\underline{1}}\}}] + \mathbb{E}_{\mathbf{c}}[\tau_{+\underline{1}}]\right)[1 + o(1)] \tag{3.23}$$

The next lemma regards the estimation of the two addenda of (3.23), using the potential theoretic approach:

Lemma 3.5. *There exist two positive constants $k_1, k_2 < \infty$ such that*

$$\frac{\mu(-\underline{1})}{\mathrm{cap}_\beta(-\underline{1}, \{\mathbf{c}, +\underline{1}\})} = \frac{e^{\beta\Gamma_{\mathrm{m}}}}{k_2}[1 + o(1)] \quad and \quad \frac{\mu(\mathbf{c})}{\mathrm{cap}_\beta(\mathbf{c}, +\underline{1})} = \frac{e^{\beta\Gamma_{\mathrm{m}}}}{k_1}[1 + o(1)]. \tag{3.24}$$

By general standard results of the potential theoretic approach, it can be shown indeed that $\mathbb{E}_{-\underline{1}}[\tau_{\mathbf{c}}\mathbf{1}_{\{\tau_{\mathbf{c}}<\tau_{+\underline{1}}\}}]$ equals the left hand side in the first of (3.24) and $\mathbb{E}_{\mathbf{c}}[\tau_{+\underline{1}}]$ equals the left hand side in the second of (3.24). Hence, by (3.23), Lemmata 3.2, 3.3, 3.4 and 3.5, Theorem 3.1 follows.

4   Sketch of proof of Theorem 3.1

In this section we prove Theorem 3.1, by proving Lemmata 3.2, 3.3, 3.4, and 3.5.

Due to space constraints we sketch the main idea behind the proof of Lemmata 3.2, 3.3, 3.4 and we give in detail the proof of 3.5. As regards Lemma 3.2, the energy inequalities and $X^s = \{+\underline{1}\}$, easily follow by (2.7). However, in order to prove $X^m = \{-\underline{1}, \mathbf{c}\}$, for any $\sigma \in \mathcal{S} \setminus (X^s \cup X^m)$, we have to show that there exists a path $\omega : \sigma \to \mathcal{I}_\sigma$ such that the maximal *communication height* to overcome to reach a configuration at lower energy is smaller than $\Gamma_m + E(\sigma)$, i.e. $\Phi_\omega < \Gamma_m + E(\sigma)$. By Property 3.3 of Ref. [2] for all configurations σ there exists a downhill path to a configuration consisting of union of rectangular droplets: for instance rectangular checkerboard in a see of minuses or well separated plus droplets in a sea of minuses or inside a checkerboard droplet. In case these droplets are non-interacting (i.e. at distance larger than one), by the analysis of the growth/shrinkage mechanism of rectangular droplets contained in [2], it is straightforward to find the required path. In case of interacting rectangular droplets a more accurate analysis is required, but this is outside the scope of the present paper. Lemma 3.3 is a consequence of the *exit tube* results contained in [2], while Lemma 3.4 follows by Theorem 3.1 and (3.7) in [12] with an appropriate constant.

4.1   Proof of Lemma 3.5

We recall the definition of the cycle $\mathcal{A}_{-\underline{1}}$ playing the role of a generalized *basin of attraction* of the $-\underline{1}$ phase:

$$\mathcal{A}_{-\underline{1}} := \{\eta \in \mathcal{G}_{-\underline{1}} : \exists \omega = \{\omega_0 = \eta, ..., \omega_n = -\underline{1}\} \text{ such that } \omega_0, ..., \omega_n \in \mathcal{G}_{-\underline{1}}$$
$$\text{and } \Phi_\omega < \Gamma + E(-\underline{1})\}$$

where $\mathcal{G}_{-\underline{1}}$ is the set defined in [Sect. 4, [2]], containing the *sub-critical* configurations (e.g. a single checkerboard rectangle in a see of minuses, with shortest side smaller than the critical length λ). In a very similar way, we can define

$$\mathcal{A}_{\{+\underline{1},\mathbf{c}\}} := \{\eta \in \mathcal{G}^c_{-\underline{1}} : \exists \omega = \{\omega_0 = \eta, ..., \omega_n \in \{\mathbf{c}, +\underline{1}\}\} \text{ such that}$$
$$\omega_0, ..., \omega_n \in \mathcal{G}^c_{-\underline{1}} \text{ and } \Phi_\omega < \Gamma + E(-\underline{1})\}$$

We start proving the equality on the left in (3.24) by giving an upper and lower estimate for the capacity $\text{cap}_\beta(-\underline{1}, \{+\underline{1}, \mathbf{c}\})$. Thus, what we need is the precise estimates on capacities, via sharp upper and lower bounds.

Usually the upper bound is the simplest because it can be given by choosing a suitable test function. Instead, for the lower bound, we use the monotonicity of the Dirichlet form in the transition probabilities via simplified processes. Therefore, we firstly identify the domain where f^* is close to one and to zero, in our case the set $\mathcal{A}_{-\underline{1}}$ and $\mathcal{A}_{\{+\underline{1},\mathbf{c}\}}$ respectively. Restricting the processes on these sets and by rough estimates on capacities we are able to give a sharper lower bound for the capacities themselves.

Upper bound. We use the general strategy to prove an upper bound by guessing some a priori properties of the minimizer, f^*, and then to find the minimizers within this class. Let us consider the two basins of attraction $\mathcal{A}_{-\underline{1}}$ and $\mathcal{A}_{-\underline{1},\{\mathbf{c},+\underline{1}\}}$. A potential f^u will provide an upper bound for the capacity, i.e. the Dirichlet form evaluated at the equilibrium potential $f^*_{-\underline{1},\{\mathbf{c},+\underline{1}\}}$, solution of the variational problem (3.18), where the two sets with the boundary conditions are $-\underline{1}$ and $\{c, +\underline{1}\}$. We choose the following test function for giving an upper bound for the capacity:

$$f^u(x) := \begin{cases} 1 & x \in \mathcal{A}_{-\underline{1}}, \\ 0 & x \in \mathcal{A}_{-\underline{1}}^c \end{cases} \tag{4.25}$$

so that:

$$\mathcal{E}(f^u) = \frac{1}{Z} \sum_{\substack{\sigma \in \mathcal{A}_{-\underline{1}}, \\ \eta \in \mathcal{A}_{\{c,+\underline{1}\}}}} e^{-\beta H(\sigma, \eta)} + \frac{1}{Z} \sum_{\substack{\sigma \in \mathcal{A}_{-\underline{1}}, \\ \eta \in (\mathcal{A}_{\{c,+\underline{1}\}} \cup \mathcal{A}_{-\underline{1}})^c}} e^{-\beta H(\sigma, \eta)} \tag{4.26}$$

Therefore, by Lemma 4.1 in [2], (4.26) can be easily bounded by:

$$\mathcal{E}(h^u)/\mu(-\underline{1}) \leq K\, e^{-\beta \Gamma} + |\mathcal{S}|\, e^{-\beta(\Gamma + \delta)}, \tag{4.27}$$

where $\delta > 0$, because $\Gamma + E(-\underline{1})$ is nothing but the minmax between $\mathcal{A}_{-\underline{1}}$ and its complement and $E(\sigma, \eta) = \Gamma + E(-\underline{1})$ only in the transition between configurations belonging to $\mathcal{P}' \subset \mathcal{A}_{-\underline{1}}$, and some particular configurations belonging to $\mathcal{P} \subset \mathcal{A}_{\{c,+\underline{1}\}}$ for all the other transitions we have $E(\sigma, \eta) > \Gamma + E(-\underline{1})$. In particular, $\mathcal{P}'$ is the set of configurations consisting of rectangular checkerboard $R_{\lambda, \lambda-1}$ of sides λ and $\lambda - 1$ in a see of minuses. $\mathcal{P}$ is instead the subset of *critical configurations* obtained by flipping a single site adjacent to a plus spin of the internal checkerboard along the larger side of a configuration $\eta \in \mathcal{P}'$.

Lower bound. In order to have a lower bound, let us estimate the equilibrium potential. We can prove the following Lemma:

Lemma 4.6. $\exists C, \delta > 0$ such that for all $\beta > 0$

$$\min_{\eta \in \mathcal{A}_{-\underline{1}}} f^\star(\eta) \geq 1 - Ce^{-\delta\beta} \quad and \quad \max_{\eta \in \mathcal{A}_{\{c,+\underline{1}\}}} f^\star(\eta) \leq Ce^{-\delta\beta} \tag{4.28}$$

Proof. Using a standard renewal argument, given $\eta \notin \{-\underline{1}, c, +\underline{1}\}$:

$$\mathbb{P}_\eta(\tau_{\{c,+\underline{1}\}} < \tau_{-\underline{1}}) = \frac{\mathbb{P}_\eta(\tau_{\{c,+\underline{1}\}} < \tau_{-\underline{1} \cup \eta})}{1 - \mathbb{P}_\eta(\tau_{-\underline{1} \cup \{c,+\underline{1}\}} > \tau_\eta)}$$

and

$$\mathbb{P}_\eta(\tau_{-\underline{1}} < \tau_{\{c,+\underline{1}\}}) = \frac{\mathbb{P}_\eta(\tau_{-\underline{1}} < \tau_{\{c,+\underline{1}\} \cup \eta})}{1 - \mathbb{P}_\eta(\tau_{-\underline{1} \cup \{c,+\underline{1}\}} > \tau_\eta)}.$$

If the process started at point η wants to realize indeed the event $\{\tau_{\{c,+\underline{1}\}} < \tau_{-\underline{1}}\}$ it can either go to $-\underline{1}$ immediately and without returning to η again, or it may return to η without going to $\{c, +\underline{1}\}$ or $-\underline{1}$. Clearly, once the process returns to η, we can use the strong Markov property. Thus

$$\mathbb{P}_\eta(\tau_{\{c,+\underline{1}\}} < \tau_{-\underline{1}}) = \mathbb{P}_\eta(\tau_{\{c,+\underline{1}\}} < \tau_{-\underline{1} \cup \eta}) + \mathbb{P}_\eta(\tau_\eta < \tau_{\{c,+\underline{1}\} \cup -\underline{1}} \wedge \tau_{\{c,+\underline{1}\}} < \tau_{-\underline{1}})$$
$$= \mathbb{P}_\eta(\tau_{\{c,+\underline{1}\}} < \tau_{-\underline{1} \cup \eta})$$
$$+ \mathbb{P}_\eta(\tau_\eta < \tau_{\{c,+\underline{1}\} \cup -\underline{1}})\mathbb{P}_\eta(\tau_{\{c,+\underline{1}\}} < \tau_{-\underline{1}})$$

and, solving the equation for $\mathbb{P}_\eta(\tau_{\{\mathbf{c},+\underline{1}\}} < \tau_{-\underline{1}})$, we have the renewal equation. Then $\forall \eta \in \mathcal{A}_{-\underline{1}} \setminus \{-\underline{1}\}$ we have:

$$f^\star(\eta) = 1 - \mathbb{P}_\eta(\tau_{\{\mathbf{c},+\underline{1}\}} < \tau_{-\underline{1}}\}) = 1 - \frac{\mathbb{P}_\eta(\tau_{\{\mathbf{c},+\underline{1}\}} < \tau_{-\underline{1} \cup \eta})}{\mathbb{P}_\eta(\tau_{-\underline{1} \cup \{\mathbf{c},+\underline{1}\}} < \tau_\eta)}$$

and, hence,

$$f^\star(\eta) \geq 1 - \frac{\mathbb{P}_\eta(\tau_{\{\mathbf{c},+\underline{1}\}} < \tau_\eta)}{\mathbb{P}_\eta(\tau_{-\underline{1}} < \tau_\eta)}$$

For the last term we have the equality:

$$\frac{\mathbb{P}_\eta(\tau_{\{\mathbf{c},+\underline{1}\}} < \tau_\eta)}{\mathbb{P}_\eta(\tau_{-\underline{1}} < \tau_\eta)} = \frac{\mathrm{cap}_\beta(\eta, \{\mathbf{c},+\underline{1}\})}{\mathrm{cap}_\beta(\eta, -\underline{1})} \tag{4.29}$$

The upper bound for the numerator of (4.29) is easily obtained through the upper bound on $\mathrm{cap}_\beta(-\underline{1}, \{\mathbf{c},+\underline{1}\})$ which we already have. The lower bound on the denominator is obtained by reducing the state space to a single path from η to $-\underline{1}$, picking an *optimal* path $\omega = \{\omega_0, \omega_1, ..., \omega_N\}$ that realizes the minmax $\Phi(\eta, -\underline{1})$ and ignoring all the transitions that are not in the path. Indeed by Theorem 3.2, we use the monotonicity of the Dirichlet form in the transition probabilities $p(\sigma, \eta)$, for $\sigma \neq \eta$. Thus, we can have a lower bound for capacities by simply setting some of the transition probabilities $p(\sigma, \eta)$ equal to zero. It is clear that if enough of these are set to zero, we obtain a chain where everything can be computed easily. With our choice we have:

$$\mathrm{cap}_\beta(\eta, -\underline{1}) \geq \min_{\substack{f:\omega \to [0,1] \\ f(\omega_0)=1, f(\omega_N)=0}} \mathcal{E}^\omega(f) \tag{4.30}$$

where the Dirichlet form $\mathcal{E}^\omega(f)$ is defined as $\mathcal{E}$ in (3.16), with $\mathcal{S}$ replaced by ω. Due to the one-dimensional nature of the set ω, the variational problem in the right hand side can be solved explicitly by elementary computations. One finds that the minimum equals

$$M = \left[\sum_{k=0}^{N-1} Z \, e^{\beta H(\omega_k, \omega_{k+1})} \right]^{-1} \tag{4.31}$$

and it is uniquely attained at f given by

$$f(\omega_k) = M \sum_{l=0}^{k-1} Z \, e^{\beta H(\omega_l, \omega_{l+1})} \qquad k = 0, 1, ..., N. \tag{4.32}$$

Therefore,

$$\mathrm{cap}_\beta(\eta, -\underline{1}) \geq M \geq \frac{1}{KZ} \max_k e^{-\beta H(\omega_k, \omega_{k+1})}$$

and hence

$$\text{cap}_\beta(\eta, -\underline{1}) \geq C_1 \frac{1}{Z} e^{-\beta \Phi(\eta, -\underline{1})}$$

with $\lim_{\beta \to \infty} C_1 = 1/K$. Moreover, we know that if $\eta \in \mathcal{A}_{-\underline{1}}$ then it holds $\Phi(\eta, -\underline{1}) < \Phi(\eta, \{\mathbf{c}, +\underline{1}\})$. Indeed, by the definition of the set $\mathcal{A}_{-\underline{1}}$:

$$\Phi(\eta, \{\mathbf{c}, +\underline{1}\}) \geq \Gamma + E(-\underline{1}) > \Phi(\eta, -\underline{1}). \tag{4.33}$$

For this reason

$$f^\star(\eta) \geq 1 - C(\eta) e^{-\beta(\Phi(\eta, \{\mathbf{c}, +\underline{1}\}) - \Phi(\eta, -\underline{1}))} \geq 1 - C(\eta) e^{-\beta\delta},$$

and we can take $C := \sup_{\eta \in \mathcal{A}_{-\underline{1}} \setminus -\underline{1}} C(\eta)$. Otherwise $\forall \eta \in \mathcal{A}_{\{\mathbf{c}, +\underline{1}\}} \setminus \{\mathbf{c}, +\underline{1}\}$ we have:

$$f^\star(\eta) = \mathbb{P}_\eta(\tau_{-\underline{1}} < \tau_{\{\mathbf{c}, +\underline{1}\}}) = \frac{\mathbb{P}_\eta(\tau_{-\underline{1}} < \tau_{\{\mathbf{c}, +\underline{1}\} \cup \eta})}{\mathbb{P}_\eta(\tau_{-\underline{1} \cup \{\mathbf{c}, +\underline{1}\}} < \tau_\eta)}$$

and hence

$$f^\star(\eta) \leq \frac{\mathbb{P}_\eta(\tau_{-\underline{1}} < \tau_\eta)}{\mathbb{P}_\eta(\tau_{\{\mathbf{c}, +\underline{1}\}} < \tau_\eta)} = \frac{\text{cap}_\beta(\eta, -\underline{1})}{\text{cap}_\beta(\eta, \{\mathbf{c}, +\underline{1}\})} \leq C(\eta) e^{-\beta\delta}$$

proving the second equality (4.28) with $C = \max_{\eta \in \mathcal{A}_{\{\mathbf{c}, +\underline{1}\}} \setminus \{\mathbf{c}, +\underline{1}\}} C(\eta)$. $\qquad \square$

Now we are able to give a lower bound for the capacity. By (3.18), we have:

$$\text{cap}_\beta(-\underline{1}, \{\mathbf{c}, +\underline{1}\}) = \mathcal{E}(f^\star) \geq \sum_{\substack{\sigma \in \mathcal{A}_{-\underline{1}} \\ \eta \in \mathcal{A}_{\{\mathbf{c}, +\underline{1}\}}}} \mu(\sigma) p(\sigma, \eta) (f^\star(\sigma) - f^\star(\eta))^2$$

$$\geq K \mu(-\underline{1}) e^{-\beta\Gamma} + o(e^{-\beta\delta})$$

Now we want to evaluate the combinatorial pre-factor K of the sharp estimate. We have to determinate all the possible ways to choose a critical droplet in the lattice with periodic boundary conditions. We know that the set $\mathcal{P}$ of such configurations contains all the checkerboard rectangles $R_{\lambda-1, \lambda}$ in a see of minuses (see Fig. 5).

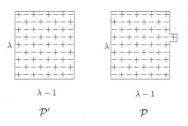

$$\mathcal{P}' \qquad\qquad \mathcal{P}$$

Fig. 4. *Saddles configurations.*

Because of the translational invariance on the lattice, we can associate at each site x two rectangular droplets $R_{\lambda-1,\lambda}$ and $R_{\lambda,\lambda-1}$ such that their north-west corner is in x. Considering the periodic boundary conditions and being Λ a square of side L, the number of such rectangles is $A = 2L^2$ In order to calculate K, we have to count in how many ways we can add a protuberance to a rectangular checkerboard configuration $R_{\lambda-1,\lambda}$, along the largest side and adjacent to a plus spin of the checkerboard. Hence, we have that $K = 4A\lambda = 8\lambda L^2$, and this completes the proof of the first equality in (3.24). The proof of the second equality in (3.24) can be achieved using very similar arguments. (Fig. 4) $\square$

Acknowledgements. The authors thank A. Bovier, F. den Hollander, M. Slowick, and A. Gaudillière for valuable discussions.

References

1. Bovier, A., Eckhoff, M., Gayrard, V., Klein, M.: Metastability and low lying spectra in reversible Markov chains. Comm. Math. Phys. **228**, 219–255 (2002)
2. Cirillo, E.N.M., Nardi, F.R.: Metastability for the ising model with a parallel dynamics. J. Stat. Phys. **110**, 183–217 (2003)
3. Cirillo, E.N.M., Nardi, F.R., Spitoni, C.: Metastability for reversible probabilistic cellular automata with self-interaction. J. Stat. Phys. **132**, 431–471 (2008)
4. Cirillo, E.N.M., Nardi, F.R., Spitoni, C.: Competitive nucleation in reversible probabilistic cellular automata. Phys. Rev. E **78**, 040601 (2008)
5. Cirillo, E.N.M., Nardi, F.R., Spitoni, C.: Competitive nucleation in metastable systems. Applied and Industrial Mathematics in Italy III, Series on Advances in Mathematics for Applied Sciences, vol. 82, pp. 208-219 (2010)
6. Cirillo, E.N.M., Nardi, F.R., Spitoni, C.: Sum of exit times in series of metastable states. Preprint 2016. arXiv:1603.03483
7. Grinstein, G., Jayaprakash, C., He, Y.: Statistical mechanics of probabilistic cellular automata. Phys. Rev. Lett. **55**, 2527–2530 (1985)
8. Kari, J.: Theory of cellular automata: a survey. Theor. Comput. Sci. **334**(1–3), 3–33 (2005)
9. Landim, C., Lemire, P.: Metastability of the two-dimensional Blume-Capel model with zero chemical potential, small magnetic field. Preprint 2016. arXiv:1512.09286
10. Lebowitz, J.L., Maes, C., Speer, E.: Statistical mechanics of probabilistic cellular automata. J. Stat. Phys. **59**, 117–170 (1990)
11. Louis, P.-Y.: Effective parallelism rate by reversible PCA dynamics. In: Wąs, J., Sirakoulis, G.C., Bandini, S. (eds.) ACRI 2014. LNCS, vol. 8751, pp. 576–585. Springer, Heidelberg (2014)
12. Manzo, F., Nardi, F.R., Olivieri, E., Scoppola, E.: On the essential features of metastability: tunnelling time and critical configurations. J. Stat. Phys. **115**, 591–642 (2004)
13. Mairesse, J., Marcovici, I.: Around probabilistic cellular automata. Theor. Comput. Sci. **559**, 42–72 (2014)
14. Nardi, F.R., Spitoni, C.: Sharp asymptotics for stochastic dynamics with parallel updating rule. J. Stat. Phys. **146**, 701–718 (2012)
15. Olivieri, E., Scoppola, E.: Markov chains with exponentially small transition probabilities: first exit problem from a general domain. I. the reversible case. J. Stat. Phys. **79**, 613–647 (1995)

16. Olivieri, E., Vares, M.E.: Large Deviations and Metastability. Cambridge University Press, UK (2004)
17. Bandini, S., Sirakoulis, G.C. (eds.): ACRI 2012. LNCS, vol. 7495. Springer, Heidelberg (2012)

Partial Reversibility of One-Dimensional Cellular Automata

Ronaldo de Castro Corrêa[2(✉)] and Pedro P.B. de Oliveira[1,2(✉)]

[1] Faculdade de Computação e Informática, Universidade Presbiteriana Mackenzie,
Rua da Consolação, 930 - Consolação, São Paulo, SP 01302-907, Brazil
[2] Pós-Graduação em Engenharia Elétrica e Computação,
Universidade Presbiteriana Mackenzie, Rua da Consolação, 930 -
Consolação, São Paulo, SP 01302-907, Brazil
ronaldo.c.correa@gmail.com, pedrob@mackenzie.br

Abstract. Reversibility is the property of very special cellular automata rules by which any path traversed in the configuration space can be traversed back by its inverse rule. Expanding this context, the notion of partial reversibility has been previously proposed in the literature, as an attempt to refer to rules as being more or less reversible than others, since some of the paths of non-reversible rules could be traversed back. The approach was couched in terms of a characterisation of the rule's pre-image pattern, that is, the number of pre-images of a rule for all configurations up to a given size, and their relative lexicographical ordering used to classify the rules in terms of their relative partial reversibility. Here, we reassess the original definition and define a measure that represents the reversibility degree of the rules, also based on their pre-image patterns, but now relying on the probability of correctly reverting each possible cyclic, finite length configuration, up to a maximum size. As a consequence, it becomes possible to look at partial reversibility in absolute terms, and not relatively to other rules, as well to infer the reversibility degrees for arbitrary lattice sizes, even in its limit to infinity. All the discussions are restricted to the elementary space, but are also applicable to any one-dimensional rule space.

Keywords: One-dimensional cellular automata · Reversible rule · Partial reversibility · Reversibility degree · Elementary space · Pre-image pattern

1 Introduction

The rules of some cellular automata (CAs) permit that any temporal evolution be reversed, regardless of the initial configuration, by means of its corresponding inverse rule. This property of reversibility is appealing due to the various potential applications it may entail, such as in encryption, reversible computing processes, quantum computation, etc. [10].

© IFIP International Federation for Information Processing 2016
Published by Springer International Publishing Switzerland 2016. All Rights Reserved
M. Cook and T. Neary (Eds.): AUTOMATA 2016, LNCS 9664, pp. 120–134, 2016.
DOI: 10.1007/978-3-319-39300-1_10

Several studies on the reversibility of cellular automata have been carried out, both aiming at understanding the properties of these rules, and at creating algorithms to detect or enumerate reversible cellular automata. For example, [1,6] propose algorithms to build reversible rules, the former using a method based on graphs and the latter based on an elaborate algebraic approach. Algorithms also do exist to establish whether a one-dimensional rule is reversible or not, such as in [5], since this problem is undecidable for larger dimensions [4]. Also, there are simple and effective techniques to build reversible CAs using variants of the CAs. One of them are the block CAs, where groups of cells have their states changed together [9]; another are the partitioned celllular automata (PCAs), introduced in [8], which can be regarded as a CA with multiple tracks; and finally, the second order CAs introduced in [11], exemplified in the space of elementary rules. The facility to build and handle reversible CAs with these techniques has been important both in expanding the phenomenology of reversible CAs, and in proving important theoretical results about (conventional) reversible CAs. As such, [7] used PCAs to show that one-dimensional reversible CAs can be computation universal, and [3] proved that any reversible CA can be represented by a reversible block CA.

Broadening the original reversibility concept, [2] introduced the notion of *relative partial reversibility*, which consists in classifying one-dimensional rules that are more or less reversible, clustering them if they have the same partial reversibility. Since reversible CAs can be very difficult to find in an arbitrary space, a practical motivation for defining ways to characterise partial reversibility is the possibility of iteratively searching that space, following the path of rules which are more and more reversible than others, until reaching the reversible ones, much like the successful approach discussed in [13]. However, exploring partial reversibility is tempting in itself, as a way to probe a rule space from the perspective of the reversibility property.

Under the scheme presented in [2], the set of reversible rules are placed at the top of the rank, and the less reversible ones at the bottom. Such a classification is obtained by means of the relative lexicographical ordering of the *pre-image patterns* of the rules, which is a multiset of the pre-image quantities of all possible cyclic initial configurations, up to a maximum given lattice size L_{max}.

Here, after verifying that the lexicographical ordering of pre-image patterns is not fully adequate as a basis for addressing partial reversibility, we go on defining a quantity that stands for the reversibility degree of a rule and then use it to individually analyse all rules of the elementary space. In spite of this restriction, any other one-dimensional space is also amenable to analogous analyses, provided sufficient computational resources are available. Whether our study can be used to address spaces with higher dimensionality is an open question that we do not venture to address; however, given the impossibility of an algorithm existing to verify the reversibility of rules with dimensions larger than 1 [4], we suspect that our approach may not be readily applicable. Finally, the fact that we only address cyclic configurations has no theoretical consequences, since the global function of any one-dimensional cellular automaton is bijective, if and only if, its restriction on periodic configurations is also bijective (see, for instance, [4]).

After presenting the required background for the paper in Sect. 2, the subsequent section revisits partial reversibility, the reversibility degree of a rule is proposed, and the reversibility degrees of the rules of the elementary space are individually analysed for finite lattice sizes and as well as when size tends to infinity. The last section wraps up the work with some concluding remarks.

2   Background

2.1   Basics

Cellular automata (CAs) are discrete dynamical systems defined by the triplet (S, N, f), in which S is the finite set of states $S = \{s_0, s_1, \ldots, s_{k-1}\}$, $N \in S^m$ defines the neighbourhood of a cell c, with m being the number of cells in the neighbourhood, and $f : S^m \longrightarrow S$ is the local state transition function applied to each cell. The CA lattice is a d-dimensional array of cells, with $d > 0$, each one taking on the states in S. We denote the global lattice configuration by $C \in \mathbb{Z}^d$. The rule is applied to all cells of the lattice, synchronously, at each time step t, after which the configuration C is updated.

Here we rely upon one-dimensional CAs, with $S = \{0, 1\}$ and $m = 3$, that is, the elementary space. The state transition function f receives as parameter the states of a cell c and those of its two next-neighbour cells at time t and returns the state of c at time $t + 1$. When f is defined for all S^m possible neighbourhoods the CA rule is well-defined; in the elementary space a rule might be the following set of state transitions, in reverse lexicographical order of the neighbourhoods: $\{111 \to 0, 110 \to 1, 101 \to 1, 100 \to 0, 011 \to 0, 010 \to 1, 001 \to 1, 000 \to 0\}$. Following such an ordering, each rule can be represented by an integer R, after the resulting decimal number generated from the outputs of each neighbourhood; for the previous example, the sequence of out bits corresponds to rule number 102, out of the 256 possible rules that define the space [11].

From the standpoint of their dynamical behaviours, rules may be equivalent to others, defining equivalence classes, obtained from the operations of conjugation and reflection, and by composing the latter two [11]. For binary CAs, conjugation consists of flipping all bits in the neighbourhoods and their outputs and reordering the state transitions, as mentioned above; on its part, reflection consists of reversing the neighbourhoods, while keeping the same output bit, and then performing the reordering. The equivalence class is completed by composing conjugation and reflection in any order, and reordering the resulting state transitions. Once a class of dynamical equivalence is obtained, we can simply refer to it by a representative, usually the rule with the lowest number; for instance, elementary rule 45 is the representative of the class $\{45, 75, 101, 89\}$.

The pre-images of a configuration at time step t are formed by all the possible previous configurations at $t - 1$. For a rule to be reversible, all possible configurations can only have a single pre-image. Configurations that do not have pre-image are known as *Garden of Eden* (*GoE*). If a rule has at least one *GoE* configuration, consequently, the rule also has some configuration with more than one pre-image; therefore, the rule is not reversible [4]. Reversible cellular automata are exceptions, because the majority of the rules of any space have some GoE configuration. In the elementary space, only the rules 15, 85, 51, 170, 240 and 204 are reversible.

2.2 Partial Reversibility Classes

The notion of partial reversibility of CA rules was originally proposed in [2], based on the pre-image pattern of a rule, which is a multiset containing all pre-image quantities of all possible configurations with finite lengths varying from 1 to L_{max}. More precisely, the pre-image pattern of a rule with number R, with respect to configuration sizes up to L_{max}, is represented by $\mathcal{P}_R$ and defined as:

Definition 1 (Pre-image pattern)

$$\mathcal{P}_R = \{l_1, l_2, \ldots, l_{|\mathcal{P}_R|}\} \begin{cases} |\mathcal{P}_R| \leq |S|^{C_{L_{max}}} \\ l_i \in \mathbb{Z}^+ \text{ is the number of pre-images of configuration } i \end{cases}$$

where $C_{L_{max}}$ is the total number of possible configurations of sizes from 1 to L_{max}, given by:

$$C_{L_{max}} = \sum_{q=1}^{L_{max}} |S|^q = -2 + 2^{L_{max}+1} \tag{1}$$

In words, each $l_i \in \mathcal{P}_R$ represents the number of pre-images of every (non-GoE) configuration, with sizes in the range $[1, L_{max}]$.

Remarks

(1) Notice that $\sum_{i=1}^{|\mathcal{P}_R|} l_i = C_{L_{max}}$.

(2) Pre-image pattern was first introduced in [2], therein named reversibility pattern, with the necessity that the multiset should be ordered from the lowest to the highest value. Although this is not a necessity for present purposes, we preserve the same scheme just for the sake of clarity of presentation.

The pre-image patterns of elementary rules 1 and 68 are illustrated below, with $L_{max} = 4$:

Example 1 (Pre-image patterns of elementary rules 1 and 68).

$$\mathcal{P}_1 = \{1, 1, 1, 1, 1, 1, 1, 1^*, 1, 3, 7, 11\}$$
$$\mathcal{P}_{68} = \{1, 1, 1, 1, 1, 1, 1, 2^*, 2, 2, 2, 2, 2, 2, 2, 2, 3\}$$

Rules that share the same pre-image patterns can be grouped into classes, referred to in [2] as *Partial Reversibility Classes*, PRCs for short. In face of these classes, [2] asked how they could be classified from the least to the most reversible, and proposed this would be based on their pre-image patterns, lexicographically ordered in relation to each other. From this standpoint, for the pre-image patterns $\mathcal{P}_1$ and $\mathcal{P}_{68}$ above, rule 1 turns out to be more reversible than rule 68, since at the position highlighted with "*" for both rules – which is the first position where they differ – the number of pre-images of rule 1 is smaller than that for rule 68.

In [2], such a *relative* classification of the rules was performed in the elementary space up to $L_{max} = 23$, resulting in 45 PRCs, as shown in Table 1; later on, the same authors managed to extend the classification up to $L_{max} = 31$, with no change in the classes. With all previous data available to us, here we extended them further to $L_{max} = 32$ and once again the same classes and their order remained. Quite coherently, the first PRC {15, 51, 170, 204} is formed by the reversible rules of the elementary space, while the last PRC {0} is the least reversible, since for any lattice size, a single configuration has all the others as pre-images.

Table 1. Partial reversibility classes of the elementary space.

{15, 51, 170, 204},{45, 154},{30, 106},{105, 150},{37, 164},{22, 104},{62, 110},
{25, 152}, {41, 134},{73, 146},{26, 74},{57, 156},{94, 122},{7, 168},{58, 78},{54, 108},
{14, 42},{38, 44},{35, 140},{28, 50, 56, 76},{33, 132},{77, 178},{23, 232},{6, 40},
{9, 130},{5, 160},{27, 172},{3, 19, 136, 200},{13, 162},{18, 72},{43, 142},{29, 184},
{1, 128},{4, 32},{11, 138},{12, 34},{2, 8},{10},{60},{90},{126},{36},{24},{46},**{0}**

Notice in the table that each PRC is defined only by the set of every representative rule of each dynamical equivalence class involved; for instance, while PRC {45, 154} refers to only two rules, more rules are effectively therein, since rule 45 has the same pre-image pattern of its dynamically equivalent rules 75, 101 and 89, and rule 154 has the same pre-image pattern of its dynamical class 210, 166 and 154. The fact that we are referring only to {45, 154} is a mere simplification of the notation.

Furthermore, in order to simplify any reference to specific PRCs, throughout the paper we sometimes simply refer to the first rule of the class – its representative, with the smallest number – but meaning its entire class.

3   Reversibility Degree

3.1   Reappraisal of Relative Partial Reversibility in [2]

Although the notion of relative partial reversibility in [2] was interesting for its own sake and led to insightful considerations about the pre-image patterns of the rules with increasing lattice sizes, it is insufficient if an absolute characterisation of partial reversibility is the target. The point is that relative lexicographical ordering of the pre-image patterns ends up not using them fully, which represents the ineffective use of all lattice sizes implicit in the patterns. This can be realised with the pre-image patterns $\mathcal{P}_1$ and $\mathcal{P}_{68}$ from Example 1. The lexicographical ordering thus goes on only up to l_8 (highlighted with "*"), while the remaining pre-image quantities are simply discarded, as there is no need for them.

So, there cannot be a situation in which all pre-image patterns are compared and only the last pre-image quantity is different, because the sum of all pre-image quantities must be maintained. As a consequence, to come up with a quantity that could be used to formalise such a scheme of partial reversibility seems a fruitless effort, as it would not be related to an intrinsic property of a rule. This is why unsuccessful attempts were reported in [2] exactly with that aim. What we really need is a way to account for the pre-image pattern of a rule in its entirety and regardless of any other rule involved, that is, in absolute terms.

3.2   Absolute Partial Reversibility

The notion of an absolute partial reversibility should be just a corollary of a notion of reversibility degree of a rule that would account for all the information available in the pre-image pattern of a rule.

In order to go about that, let us first have it clear that since $l_i \in \mathcal{P}_R$ contains the pre-image quantity of a possible configuration at any time step, $1/l_i$ can be interpreted as the probability of reversing the time evolution one step, at the configuration concerned. So, computing the reversion probability of all pre-image quantities l_i yields the individual probabilities of correctly reversing all possible configurations of lattices up to L_{max} of a rule. By adding all reversion probabilities, dividing the result by the number of possible configurations up to L_{max}, and normalising the result in the interval $[0, 100]$, 0 means that a rule is the least reversible in the rule space at issue, and 100 corresponds to the reversible rules. More precisely, the reversibility degree of a rule R is defined as:

Definition 2 (Reversibility degree of a rule)

$$\delta_R(\mathcal{P}_R, L_{max}) = \left(\frac{\sum_{i=1}^{|\mathcal{P}_R|} \frac{1}{l_i}}{-2 + 2^{L_{max}+1}} \right) \times 100$$

To exemplify the calculation of δ_R, consider the pre-image patterns $\mathcal{P}_0$ and $\mathcal{P}_{15}$, up to $L_{max} = 4$:

Example 2 (δ_R for two elementary rules with extreme values)

$\mathcal{P}_0 = \{2, 4, 8, 16\}$
$\delta_0 = (\frac{1/2+1/4+1/8+1/16}{30}) \times 100 = 3.125$

$\mathcal{P}_{15} = \{1,1,1,1,1,1,1,1,1,1,1,1,1,1,1,1,1,1,1,1,1,1,1,1,1,1,1,1,1,1\}$
$\delta_{15} = (\frac{30\times1}{30}) \times 100 = 100$

Because for rule 0, a single configuration is the image of all configurations, regardless of the lattice size, its reversibility degree has to be the lowest, non zero, in the elementary space, which indeed is the case; as a consequence it is no longer possible to refer to a *non-reversible* rule in any space, in absolute terms. At the other extreme, the reversible elementary rule 15 has maximal reversibility degree of 100.

Since the sum of each $1/l_i$ impairs a direct interpretation of the resulting value in terms of probability, we verified what would happen when using multiplication. But it would turn out that neither PRC {15, 51, 170, 204} of the reversible rules nor PRC {0}, the least reversible one, would be placed in the extreme ends of the classification. Thus, the approach based on the multiplication of each $1/l_i$ does not lead to a coherent result, although it would preserve a clear interpretation of the quantity in terms of probabilities.

The reversibility degree δ_R allows us to group the rules with the same degree, and classify them in a relative way, analogously to [2]; this was carried out for the elementary rules up to $L_{max} = 32$, and the results are plotted as the series of graphics showed in Fig. 1. At the vertical axis, the PRCs are classified from the least reversible ({0}) at the bottom, to the reversible one ({15, 51, 170, 204}), with the horizontal axis displaying the maximum lattice size L_{max} at each computation. It becomes clear that, as L_{max} increases, the relative positions of the PRCs tend to settle, even though with some local oscillations.

The position of the majority of the PRCs plotted in the graphic clearly stabilises as L_{max} gets progressively larger. However, this does not occur with all PRCs; for example, while PRCs {60} and {90} – the 5th and 8th rows from top to bottom, respectively, at $L_{max} = 32$ – move away from the least reversible ones, quite interestingly, the exact opposite trend obtained out of the lexicographical ordering scheme employed in [2].

3.3   Partial Reversibility Degree of Individual Elementary Rules

With the definition of the reversibility degree of a rule, it becomes possible to visualise the individual reversibility degree of the rules in the elementary space, as shown in Fig. 2. The initial lattice size used was 6, after eliminating a transient; the computations were performed up to $L_{max} = 32$. In the figure, the vertical axis displays the reversibility degree of the rules, normalised from 0 to 100, and the horizontal axis shows the lattice sizes up to L_{max}. Overall, it becomes clear that, as L_{max} increases, the reversibility degree of several PRCs either decrease or increase smoothly, almost monotonically, while three others do not follow this trend, keeping either an oscillation or a steady value. Relevant details of the figure are discussed below.

As expected, the graphics shows that the PRC of the reversible elementary rules ({15, 51, 170, 204}) have maximal reversibility degree of 100, and that PRC {0} contains the least reversible rules. It also becomes evident the reason why PRCs {60} and {90} moved away from the least reversible ones on the relative partial reversibility graphs of Fig. 1. The explanation is that PRC {60} (at the 5th row from top to bottom, at $L_{max} = 32$), has a stable reversibility degree of 25; consequently, as the degrees of several PRCs decrease, while L_{max} increases, they get closer to the least reversible rules. So, it is not PRC {60} that moves away from the least reversible ones, but an effect of the behaviour of the others. A similar situation occurs with PRC {90} (at the 8th row from top to bottom, at $L_{max} = 32$), with the difference that PRC {90} has an apparently periodical variation in its reversibility degree.

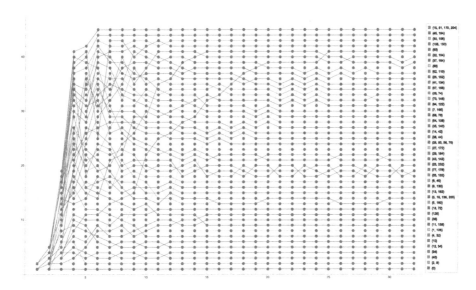

Fig. 1. Relative partial reversibility in the elementary space, as defined by the partial reversibility degrees of the rules, up to $L_{max} = 32$.

Table 2. Groups of PRCs sharing a pattern in their reversibility degrees.

$\overrightarrow{PRC}$	$\{105, 150\}, \{90\}, \{60\}$
$\uparrow PRC$	$\{15, 51, 170, 204\}, \{45, 154\}, \{30, 106\}$
$\downarrow PRC$	$\{0\}, \{2, 8\}, \{46\}, \{24\}, \{12, 34\}, \{10\}, \{4, 32\},\{1, 128\}, \{11, 138\}, \{36\}, \{126\},$ $\{18, 72\}, \{5, 160\}, \{3, 19, 136, 200\}, \{13, 162\}, \{9, 130\}, \{6, 40\}, \{33, 132\},$ $\{77, 178\}, \{23, 232\}, \{43, 142\}, \{29, 184\}, \{27, 172\}, \{28, 50, 56, 76\}, \{38, 44\},$ $\{14, 42\}, \{35, 140\},\{54, 108\},\{58, 78\}, \{7, 168\}, \{94, 122\}, \{73, 146\}, \{26, 74\},$ $\{57, 156\}, \{41, 134\}, \{25, 152\}, \{62, 110\},\{37, 164\}, \{22, 104\}$

Figure 2 is limited to $L_{max} = 32$ due to the high processing time to generate the pre-image patterns for superior lattice sizes; for this reason, it was not possible to empirically verify in which reversibility levels the PRCs would get stable. Alternatively, we went about analytically inferring the limit of the reversibility degree with L_{max} tending to infinity. For such, three groups of PRCs have been identified, shown in Table 2, as suggested by the plot, namely: $\overrightarrow{PRC}$, whose respective reversibility degrees are stable or have periodic variation; $\uparrow PRC$, whose degrees tend to 100; and $\downarrow PRC$, with reversibility degrees tending to 0.

For each group, their respective expressions for δ_R were rewritten as a function of L_{max} instead of $\mathcal{P}_R$, assuming the pre-image pattern of PRCs has a non-mutable pattern of formation, as suggested from the graphs in Fig. 2. In doing so, we managed to work out analytically the limits of the reversibility degrees, with the help of software *Mathematica*, especifically, $FindSequenceFunction[a_n]$,

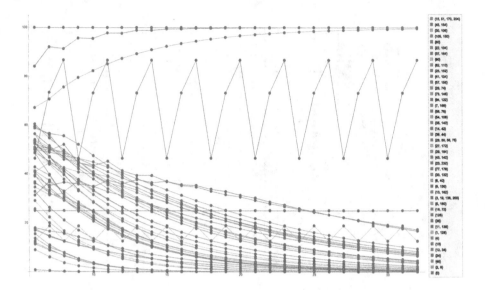

Fig. 2. Reversibility degrees of the elementary rules, with L_{max} from 6 to 32.

that tries to find a simple function that yields the sequence a_n when given successive integer arguments [12]. For all PRCs only the rule with the smallest number was used in the calculations, as a representative. Even though all PRCs have been analysed, we only discuss below significant PRCs of each group.

3.3.1   Analysis of the $\overrightarrow{PRC}$ Group

Firstly, let us look at $\overrightarrow{PRC}$, composed by $\{105, 150\}$, $\{90\}$ and $\{60\}$. It can be observed for PRC $\{105, 150\}$ that the quantities of pre-image do not change while L_{max} increases, while the frequency of pre-image quantities do (see Table 3); the pre-image pattern of rule 105 is a multiset formed by the pre-image quantities 1 and 4. As such, the frequencies of these quantities were listed as a function of L_{max}, and the series analysed with the software *Mathematica*, resulting in the following equations $q1_{105}$ and $q4_{105}$, where $x = L_{max}$:

$$q1_{105}(x) = (2/21)\left(-9 - 2^x\left(-14 + 5\cos\left(2\pi x/3\right) + \sqrt{3}\sin\left(2\pi x/3\right)\right)\right)$$
$$q4_{105}(x) = (1/42)\left(-12 + 2^x\left(7 + 5\cos\left(2\pi x/3\right) + \sqrt{3}\sin\left(2\pi x/3\right)\right)\right)$$

With equations $q1_{105}$ and $q4_{105}$, it becomes possible to rewrite equation δ_{105} as a function of L_{max} only, without the pre-image pattern $\mathcal{P}_{105}$. The reversibility degree of rule 105 as a function of L_{max} is defined by equation $\delta'_{105}(x)$ below, where the normalisation factor 100 was dropped off, as it does not influence the limit to be worked out; in fact, the notation $\delta'_R(x)$ is employed for all other analyses that will follow. In order to validate the correctness of equation δ'_{105}, it was used to generate the values of reversibility degree from $L_{max} = 16$ up to

Table 3. Frequencies of the pre-image quantities for rule 105.

x = L_{max}	1	2	3	4	5	6	7	8	9	10	11	12	13	14	15
$q1_{105}(x)$	2	6	6	22	54	54	182	438	438	1462	3510	3510	11702	28086	28086
$q4_{105}(x)$	0	0	2	2	2	18	18	18	146	146	146	1170	1170	1170	9362

$L_{max} = 32$, and compared to those empirically obtained with δ_{105} (used in Fig. 2), and they matched exactly. This same kind of verification was employed for all other results described in the paper, and they all matched.

$$\delta'_{105}(x) = (q1_{105}(x) + (1/4)q4_{105}(x))/(-2 + 2^{x+1}) \tag{2}$$

Given the hypothesis that the pattern of formation of pre-image quantities would remain the same for $L_{max} > 32$, the limit of $\delta'_{105}(x)$ as x tending to infinity can then be calculated, so as to verify at which reversibility degree the PRC {105, 150} would stabilise:

$$\lim_{x \to +\infty} \delta'_{105}(x) = \left(231 - 15\left(5\cos(2\pi x/3) + \sqrt{3}\sin(2\pi x/3)\right)\right)/336$$

In doing so, it can be verified that the outcomes of $\cos(\frac{2\pi x}{3})$ and $\sin(\frac{2\pi x}{3})$ follow a pattern: for $x \bmod 3 = 0$, their resulting values are 1 and 0, respectively; for $x \bmod 3 \neq 0$, the cosine is always $-1/2$, while the sine yields $\sqrt{3}/2$ when $x \bmod 3 = 1$ or $-\sqrt{3}/2$ when $x \bmod 3 = 2$. Since the graph corresponding to PRC {105, 150} in Fig. 2 varies periodically with three values, in the limit $\delta'_{105}(x)$ also leads to three values:

$$\lim_{x \to +\infty} \delta'_{105}(x) = \begin{cases} \text{if } x \bmod 3 = 0 : 46.4286 \\ \text{if } x \bmod 3 = 1 : 73.2142 \\ \text{if } x \bmod 3 = 2 : 86.6071 \end{cases}$$

The same general procedure has been applied to PRC {90}, after realising that its pre-image pattern is a multiset composed by pre-image quantities 2 and 4. By listing the frequencies of these pre-image quantities as a function of L_{max} (following Table 4) and inferring the series with *Mathematica*, two equations are obtained:

$$q2_{90}(x) = (1/6)(-2 + 3 \times 2^x + (-1)^{1+x}2^x)$$
$$q4_{90}(x) = (1/12)(-4 + (-2)^x + 3 \times 2^x)$$

Table 4. Frequencies of the pre-image quantities for rule 90.

x = L_{max}	1	2	3	4	5	6	7	8	9	10	11	12	13	14	15
$q2_{90}(x)$	1	1	5	5	21	21	85	85	341	341	1365	1365	5461	5461	21845
$q4_{90}(x)$	0	1	1	5	5	21	21	85	85	341	341	1365	1365	5461	5461

After rewriting equation δ_{90} as a function of L_{max}, by means of functions $q2_{90}$ and $q4_{90}$ (as below), the validation procedure described earlier was performed, with renewed success, thus yielding:

$$\delta'_{90}(x) = \left((1/2)q2_{90}(x) + (1/4)q4_{90}(x)\right)/(-2 + 2^{x+1}) \tag{3}$$

Once again, the limit of $\delta'_{90}(x)$ with x tending to infinity was calculated to determine the reversibility degree to which PRC $\{90\}$ stabilises:

$$\lim_{x \to +\infty} \delta'_{90}(x) = \begin{cases} \text{if } x \bmod 2 = 0 : 12.50 \\ \text{if } x \bmod 2 = 1 : 18.75 \end{cases}$$

Repeating the previous procedures for PRC $\{60\}$, whose reversibility degree has no variations, one can realise that its pre-image pattern is made up of only frequency 2. The frequencies of the pre-image quantities were analysed as a function of L_{max} and $\delta'_{60} = ((1/2)q2_{60}(x))/(-2 + 2^{x+1})$ was obtained. The limit with x tending to infinity was calculated and the reversibility degree stabilises in 25.

3.3.2   Analysis of the $\uparrow PRC$ Group

The next group of PRCs analysed is $\uparrow PRC$, composed by those that approach reversibility degree 100, namely, PRCs $\{45, 154\}$ and $\{30, 106\}$ (excluding the trivial case of the reversible rules). The analyses followed the very same procedures as those before. As such, the pre-image patterns of rule 45 are multisets composed by the pre-image quantities 1, 2 and 3, whose frequencies have been listed (Table 5), extended by inference, eventually leading to the following three equations:

$$q1_{45}(x) = \frac{3(\sqrt{2}-2)2^{x+3} + 6\left((\sqrt{2}-2)x - (-1)^x\right) - 28(-1)^{2x}}{12(\sqrt{2}-2)}$$
$$+ \frac{3\left(-2^{\frac{x}{2}+3}\left((2\sqrt{2}-3)(-1)^x - 1\right) + \sqrt{2}(-1)^x\right) + 20\sqrt{2}(-1)^{2x} + \sqrt{2} - 14}{12(\sqrt{2}-2)}$$

$$q2_{45}(x) = \frac{2^{\frac{x}{2}+1}\left((2\sqrt{2}-3)(-1)^x - 1\right) + 2(-1)^x + 4(-1)^{2x}}{2(\sqrt{2}-2)}$$
$$+ \frac{2\left((2\sqrt{2}-3)(-1)^{2x+1} + \sqrt{2} - 1\right)x - 3\sqrt{2}(-1)^{2x} + \sqrt{2}(-1)^{x+1} + 2}{2(\sqrt{2}-2)}$$

$$q3_{45}(x) = (1/4)(-1)^x\left((-1)^x(2x - 1) + 1\right)$$

Table 5. Frequencies of the pre-image quantities for rule 45.

x = L_max	1	2	3	4	5	6	7	8	9	10	11	12	13	14	15
q1_45(x)	2	3	11	20	52	101	229	454	966	1927	3975	7944	16136	32265	65033
q2_45(x)	0	0	0	2	2	8	8	22	22	52	52	114	114	240	240
q3_45(x)	0	1	1	2	2	3	3	4	4	5	5	6	6	7	7

With functions $q1_{45}$, $q2_{45}$ and $q3_{45}$, the expression for δ_{45} can be rewritten as a function of L_{max}, yielding

$$\delta'_{45}(x) = (q1_{45}(x) + (1/2)q2_{45}(x) + (1/3)q3_{45}(x))/(-2 + 2^{x+1}) \qquad (4)$$

with the limit $\lim_{x\to+\infty} \delta'_{45}(x) = 1$. So, PRC $\{45, 154\}$ tends to reversibility in the limit of the lattice size.

Finally, PRC $\{30, 106\}$ was also analysed following the same procedure, and three equations – this time recurrence equations – were obtained for $q1_{30}$, $q2_{30}$ and $q3_{30}$ (shown on the page), resulting in

$$\delta'_{30}(x) = (q1_{30}(x) + (1/2)q2_{30}(x) + (1/3)q3_{30}(x))/(-2 + 2^{x+1}) \qquad (5)$$

thus leading to $\lim_{x\to+\infty} \delta'_{30}(x) = 1$, which means that also these rules tend to reversibility.

$$q1_{30}(x) = -4q1_{30}(x-6) - 2q1_{30}(x-5) - 4q1_{30}(x-4) + 3q1_{30}(x-3) + 2q1_{30}(x-1),$$

$$\text{with} \quad \begin{cases} q1_{30}(1) = 0, q1_{30}(2) = 2, q1_{30}(3) = 5, \\ q1_{30}(4) = 11, q1_{30}(5) = 31, q1_{30}(6) = 72 \end{cases}$$

$$q2_{30}(x) = +3 - 2q2_{30}(x-6) - q2_{30}(x-4) + 3q2_{30}(x-3) + q2_{30}(x-1),$$

$$\text{with} \quad \begin{cases} q2_{30}(1) = 1, q2_{30}(2) = 2, q2_{30}(3) = 3, \\ q2_{30}(4) = 8, q2_{30}(5) = 14, q2_{30}(6) = 24 \end{cases}$$

$$q3_{30}(x) = \tfrac{1}{9}\left(3x + \sqrt{3}\sin\left(\tfrac{2\pi x}{3}\right) + 3\cos\left(\tfrac{2\pi x}{3}\right) - 3\right)$$

3.3.3 Analysis of the ↓PRC Group

With all PRCs analysed so far their pre-image patterns are composed by constant values of pre-image quantities, only their frequencies varying. But it so happens that for ↓PRC new pre-image quantities appear in their pre-image patterns as L_{max} increases, with progressively increasing values. This impaired the possibility of inferring the frequencies as a function of L_{max}. The pre-image pattern of PRC $\{22, 104\}$ exemplifies this situation, with the exponent of each pre-image quantity representing its frequency:

Example 3 (Pre-image pattern of rule 22)

$$\mathcal{P}_{22} = \{1^{19490}, 2^{6227}, 3^{3023}, 4^{1144}, 5^{619}, 6^{347}, 7^{303}, 8^{58}, 9^{306}, 10^{41}, 11^{16}, 12^{15}, 13^{184}, 14^{44},$$
$$16^1, 19^{53}, 20^{15}, 22^1, 27^{14}, 28^{43}, 32^1, 39^{15}, 41^{30}, 47^1, 68^1, 99^1, 145^1, 212^1, 310^1\}$$

Notwithstanding the difficulty just mentioned, the limit can still be calculated with direct analysis of δ_R. The growing order of the numerator $\sum_{i=1}^{|\mathcal{P}_R|} \frac{1}{l_i}$ of δ_R resembles a harmonic series $\sum_{x=1}^{\infty} \frac{1}{x}$, which gradually diverges as x increases; on the other hand, its denominator $-2 + 2^{L_{max}+1}$, has exponential growth (Eq. 1). The left and right hand figures in Fig. 3 illustrate the difference in growth order of a harmonic series and Definition 2, respectively.

As a consequence, the reversibility degree limit is clearly 0; further analytical validation with *Mathematica* also confirms this conclusion. All the pre-image patterns of the group ↓PRC were verified, and they are similar to that of rule 22 discussed above. Therefore, all of them will tend to 0 with L_{max} tending to infinity.

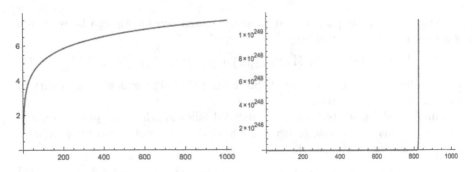

Fig. 3. Growing order of the components of δ_R in Definition 2.

4 Concluding Remarks

The original relative lexicographical ordering of pre-image patterns employed in [2] is not adequate as a way to characterise the notion of reversibility degree of a rule as an intrinsic property. We made it explicit that it rejects pre-image quantities, unless the pre-image patterns compared are identical.

We then gave the definition of the reversibility degree of a rule (δ_R) as a way to represent an absolute form of characterising the reversibility degree of a rule, within a scale from 0 to 100, from the least reversible to reversible ones. This quantity then provides a natural way to characterise the relative partial reversibility between arbitrary rules. Since δ_R relies upon the pre-image pattern of all pre-image quantities for all lattice sizes up to L_{max}, it can be regarded as a rule property up to this finite limit. In cases when δ_R can be fully written in terms of L_{max} only, and its limit to infinity calculated, the property becomes truly robust. As we showed these limits can be worked out for the entire elementary space.

Property δ_R was applied to all rules of the elementary space and the relative ordering was plotted in Fig. 1. The resulting relative ordering was similar to the one in [2], although not exactly. For example, here PRCs {60} and {90} moved away from the less reversible rules, instead of getting closer, as happened therein. It was also possible to analyse the reversibility degree of all PRCs individually as a function of L_{max}, which was not possible before.

The calculations were limited to $L_{max} = 32$ due to the high processing time and big storage space required; thus, it was not possible to empirically verify in which degree the PRCs would be stabilised. To verify the reversibility degree with L_{max} tending to infinity, the pre-image patterns were analysed through their formation patterns. The PRCs were divided into three groups to facilitate the analysis: $\overrightarrow{PRC}$, in which their respective degree are stable or have periodic variation; $\uparrow PRC$, which tend reversibility; and $\downarrow PRC$, which tend to 0.

For the groups $\overrightarrow{PRC}$ and $\uparrow PRC$, it was possible to rewrite δ_R as a function of L_{max} only (instead also of $\mathcal{P}_R$), which has permitted the calculation of their limits with L_{max} tending to infinity, thus identifying the reversible rules or those with a limit trend to reversibility. However, it was seen in the group $\downarrow PRC$ that the pre-image patterns of the PRCs do not display a formation pattern as a function of L_{max}, since new pre-image quantities appear while L_{max} increases. Nevertheless, by direct analysis of the expression that defines δ_R, it was possible to demonstrate that its reversibility levels tend to to 0 while L_{max} tend to infinity.

All discussions and analyses have been made in the context of the elementary space due to computational constraints only; any larger one-dimensional space is also amenable to the same kind of treatment. However, since we relied on numerical/computational methods, our results open various ways for formal follow-ups. Along these lines we envisage, for instance, the characterisation of the partial reversibility classes for finite lattice sizes and their limit with infinite size configurations, including the questions of why there are 45 classes for finite configurations and 5 for infinite; proofs that the series we inferred for the pre-image patterns really hold; to investigate what would be the counterpart of an inverse rule in the partial reversibility context, including the question of whether a single of multiple inverse rules should be defined; and to look at how the notion of partial reversibility in CAs extends to larger dimensions, since undecidability of reversibility in these cases is an established fact.

Acknowledgements. We thank the financial support provided by Fornax Technology to R. Corrêa, and by MackPesquisa (Fundo Mackenzie de Pesquisa), FAPESP (Proc 2005/04696-3) (Fundação de Amparo à Pesquisa do Estado de São Paulo) and CNPq (Conselho Nacional de Desenvolvimento Científico e Tecnológico) to P. de Oliveira.

References

1. Boykett, T.: Efficient exhaustive listings of reversible one dimensional cellular automata. Theor. Comput. Science. **325**, 215–247 (2004)
2. de Oliveira, P.P.B., Freitas, R.: Relative partial reversibility of elementary cellular automata. In: Kari, J., Fatés, N., Worsh, T. (eds.) Proceedings of Automata 2010: 16th International Workshop on Cellular Automata and Discrete Complex Systems, LORIA-INRIA, Nancy, France, pp. 195–208 (2010)
3. Durand-Lose, J.: Representing reversible cellular automata with reversible block cellular automata. In: Cori, R., Mazoyer, J., Morvan, M., Mosseri, R. (eds.) Discrete Models: Combinatorics, Computation, and Geometry, DM-CCG 2001. Discrete Mathematics and Theoretical Computer Science Proceedings, vol. AA, pp. 145–154 (2001)
4. Kari, J.: Theory of cellular automata: a survey. Theor. Comput. Sci. **334**, 3–33 (2005)
5. Mora, J.C.S.T., Vergara, S.V.C., Martínez, G.J., McIntosh, H.V.: Procedures for calculating reversible one-dimensional cellular automata. Physica D **202**, 134–141 (2005)
6. Moraal, H.: Graph-theoretical characterization of invertible cellular automata. Physica D **141**, 1–18 (2000)
7. Morita, K.: Computation universality of one-dimensional reversible cellular automata. Inf. Process. Lett. **42**, 325–329 (1992)

8. Morita, K., Harao, M.: Computation universality of 1 dimensional reversible (injective) cellular automata. Trans. Inst. Electron., Inf. Commun. Eng., E **72**, 758–762 (1989)
9. Toffoli, T., Margolus, N.: Cellular Automata Machines: A New Environment for Modeling. MIT Press, Cambridge (1987)
10. Toffoli, T., Margolus, N.: Invertible cellular automata: a review. Physica D **45**, 229–253 (1994)
11. Wolfram, S.: A New Kind of Science. Wolfram Media, Champaign (2002)
12. Research, W.: Wolfram Mathematica (2015). http://www.wolfram.com/mathematica
13. Wolz, D., de Oliveira, P.P.B.: Very effective evolutionary techniques for searching cellular automata rule spaces. J. Cell. Automata **3**(4), 289–312 (2008)

Two-Dimensional Traffic Rules and the Density Classification Problem

Nazim Fatès[1]([✉]), Irène Marcovici[2], and Siamak Taati[3]

[1] Inria Nancy – Grand Est, LORIA UMR 7503, Nancy, France
nazim.fates@loria.fr
[2] Institut Élie Cartan de Lorraine, Université de Lorraine, Nancy, France
irene.marcovici@univ-lorraine.fr
[3] Mathematics Institute, Leiden University, Leiden, The Netherlands
siamak.taati@gmail.com

Abstract. The density classification problem is the computational problem of finding the majority in a given array of votes, in a distributed fashion. It is known that no cellular automaton rule with binary alphabet can solve the density classification problem. On the other hand, it was shown that a probabilistic mixture of the traffic rule and the majority rule solves the one-dimensional problem correctly with a probability arbitrarily close to one. We investigate the possibility of a similar approach in two dimensions. We show that in two dimensions, the particle spacing problem, which is solved in one dimension by the traffic rule, has no cellular automaton solution. However, we propose exact and randomized solutions via interacting particle systems. We assess the performance of our models using numeric simulations.

Keywords: Density classification problem · Spacing problem · Interacting particle systems

1 Introduction

Let us imagine a medium composed of a great number of cells arranged regularly on a grid. Each cell is linked with its immediate neighbours and the only thing it can do is to change its own state according to the state of its neighbours. Can we compute with such a medium? And what happens if the updates occur at random times? And what if the cells are subject to noise?

In order to study this robustness mechanisms on a mathematical basis, we will here focus on two simple computational problems. The first problem is the *density classification problem*, which is the problem of finding the majority state in a distributed fashion. In the original setting, the computational protocol is required to be local and parallel, and use no extra memory other than the evolving configuration itself. The protocol is also required to be scalable, which means

S. Taati—The work of Siamak Taati is supported by ERC Advanced Grant 267356-VARIS of Frank den Hollander.

M. Cook and T. Neary (Eds.): AUTOMATA 2016, LNCS 9664, pp. 135–148, 2016.
DOI: 10.1007/978-3-319-39300-1_11

it must perform the task on configurations of arbitrary size. In other words, we look for a cellular automaton rule that performs the task. In this paper, we also consider variants of this problem in which the process is allowed to be asynchronous, non-deterministic or random.

This problem has attracted a considerable amount of attention these last years. It is trivial in most settings but it is not easy to solve in the case of cellular automata. The difficulty comes from the necessity to reach a consensus on the state of the cells: the system should converge to a situation with all 1s or all 0s, depending on whether the initial state contains more 0s or more 1s, respectively.

Inspired by the work of Gàcs, Kurdiumov and Levin, in 1988, Packard formulated this problem as a challenge to study genetic algorithms [11]. This triggered a wide competition to find rules with an increasing quality of classification. In 1995, Land and Belew proved that no perfect solution exists for one-dimensional deterministic systems [9]. Recently, this fact was re-demonstrated with a simpler argument and the proof was extended to probabilistic rules and to any dimension [2]. It was even shown that for any candidate solution there are configurations with a density close to 0 and 1 that are misclassified [8].

Since then, different variants of the problem have been proposed and it has been shown by various authors that relaxing one of the conditions of the problem is often sufficient to find perfect solutions [3]. In particular, Fukś proposed to combine two rules sequentially to obtain a perfect solution, see Ref. [7] and references therein. Probabilistic cellular automata could provide another interesting framework: it was discovered that although no perfect rule exists, it is possible to find a family of one-dimensional nearest-neighbour rules for which the probability of making an error of classification can be made as small as wanted [5]. The perfect solution can thus be approximated – but not reached ! – at the cost of an increase in the average time to reach a consensus.

The construction proposed for building this family of rules consists of mixing stochastically two well-known rules: the traffic rule, which introduces space between particles, and the majority rule, which has a "homogenising" effect. In this text, we ask whether there also exist a "close-to-perfect" solution for two-dimensional cellular automata. At first sight, one does not see why the problem should be significantly different for two-dimensional systems. However, there is no such thing as a "traffic" rule in two dimensions (2D). If we decompose a 2D grid in layers and apply a classical traffic rule on each layer, then different consensuses might be attained and there is no obvious means on how one can obtain the "right" global consensus from a collection of local consensuses.

We call the problem that is solved by the traffic rule in one dimension the *particle spacing problem*. We tackle this problem in two dimensions. We then (partially) solve the density classification problem by combining our particle spacing model with a local majority rule.

The outline of the article is as follows. After presenting the basic definitions and properties of our models in Sect. 2, we show the advantage of using interacting particle systems to tackle the problem (Sect. 3). We then present a concrete solution and analyse its behaviour with numerical simulations in Sect. 4.

2   Basics

2.1   Setting

In dimension $d \geq 1$, we set the cellular space to be a grid with periodic boundary conditions, defined by $\mathcal{L} = (\mathbb{Z}/n_1\mathbb{Z}) \times \cdots \times (\mathbb{Z}/n_d\mathbb{Z})$, for some $n_1, \ldots, n_d \geq 1$. The number of cells of $\mathcal{L}$ is $N_{\mathcal{L}} = n_1 \cdots n_d$. We say that the grid is *even-sized* if $n_1, \ldots, n_d$ are all even.

Each cell of this space can hold a binary state, so that the set of states is denoted by $Q = \{0, 1\}$.

The set of configurations is denoted by $\mathcal{E} = \{0, 1\}^{\mathcal{L}}$.

For a configuration $x \in \mathcal{E}$, and a state $q \in Q$, we define the density of state q by: $d_q(x) = \frac{1}{N_{\mathcal{L}}} \operatorname{Card}\{i \in \mathcal{L} \, ; \, x_i = q\}$.

For a given configuration $x \in \mathcal{E}$, we say that cell i is *isolated* if none of its adjacent cells is in state x_i.

For $q \in Q$, we say that a configuration $x \in \mathcal{E}$ is a *q-archipelago* if all the cells in state q are isolated, i.e., if x does not contain two adjacent cells in state q. We denote by $\mathcal{A}_q$ the set of q-archipelagos. In particular, if $x \in \mathcal{A}_q$, then $d_q(x) \leq 1/2$.

We also introduce $\mathcal{A} = \mathcal{A}_0 \cup \mathcal{A}_1$, the set of all archipelagos.

Fig. 1. Transformation of a random initial condition. Under some conditions (see Sect. 4 p. 10), the system will most probably converge to $1^{\mathcal{L}}$, which is a correct classification since the initial density is greater than $1/2$.

2.2   Presentation of the Problem

Recall that in this paper, we study two computational problems. The *density classification task* is the task of transforming a given configuration $x \in Q^{\mathcal{L}}$ into one of the two uniform configurations $1^{\mathcal{L}}$ or $0^{\mathcal{L}}$ depending on which of 1 or 0 has strict majority in x (see Fig. 1). More specifically, given an input $x \in Q^{\mathcal{L}}$, a computational process performing the density classification task must return $1^{\mathcal{L}}$ if $d_1(x) > 1/2$ and $0^{\mathcal{L}}$ if $d_0(x) > 1/2$. (Generally, the case $d_1(x) = d_0(x)$ is avoided.)

Our approach to solve the density classification problem is via another problem which we call the particle spacing problem. The *particle spacing problem* is the computational problem of rearranging the "particles" (say, symbols 1) on a configuration $x \in Q^{\mathcal{L}}$ so as to obtain an archipelago configuration (see Fig. 2). Again we require the computational process to be local and scalable, but we also require it to be *conservative*: at every step of the process, the number of particles (symbols 1) must be preserved.

There are two possible variants for the latter problem. In the *strict* spacing problem, we require that the sets $\mathcal{A}_1$ and $\mathcal{A}_0$ are *absorbing*, in the sense that as soon as the process enters $\mathcal{A}_q$ (for either $q = 1$ or $q = 0$), it cannot leave it. In the *loose* variant of the problem, we require the process to eventually remain in $\mathcal{A}_q$. Note that the latter is equivalent to the condition that the computation reaches an absorbing subset of $\mathcal{A}_1$ or $\mathcal{A}_0$.

For both problems, our purpose is to build a solution with a cellular system; this means that we have a set of interacting components, which can have a deterministic or stochastic behaviour, and interact only locally. We will here consider cellular automata (deterministic or stochastic) and interacting particle systems.

2.3   Known Results with Cellular Automata

We now present the principal known results concerning solutions of the density classification problem and of the particle spacing problem, using cellular automata.

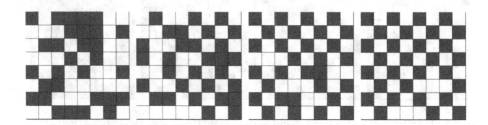

Fig. 2. Transformation of a random initial configuration into an archipelago.

A cellular automaton $F : \mathcal{E} \to \mathcal{E}$ is defined by a neighbourhood $\mathcal{N} = (v_1, \ldots, v_k) \in \mathcal{L}^k$, and by a local rule $f : Q^k \to Q$, which defines the global rule F, mapping a configuration x to the configuration $F(x)$ defined by:

$$\forall c \in \mathcal{L}, F(x)_c = f(x_{c+v_1}, \ldots, x_{c+v_k}).$$

Proposition 1. *For any $d \geq 1$, there is no deterministic cellular automaton solving the density classification problem. For $d = 1$, this means that there is no local rule f such that for any $n \geq 1$, and any $x \in \{0,1\}^{\mathbb{Z}/n\mathbb{Z}}$,*

$$d_q(x) > 1/2 \implies \exists T \geq 0, \forall t \geq T, F^t(x) = q^{\mathbb{Z}/n\mathbb{Z}}.$$

The first proof was given by Land and Belew in 1995 for dimension $d = 1$ [9]. A simplified proof was proposed in 2013 for any dimension $d \geq 1$ [2]. These results apply to deterministic cellular automata. One can ask whether stochastic transition rules could help to solve the problem. Using the same argument as for deterministic cellular automata, one can prove that there are no probabilistic cellular automata solving perfectly the density classification problem [2].

However, for $d = 1$, Fatès has provided a family of probabilistic cellular automata solving the density classification problem with an arbitrary precision [5]. This means that the probability of making a bad classification can be reduced to as low as necessary, at the cost of an increase of the average time of convergence to the uniform configuration.

The family of rules is defined with a real parameter $\epsilon > 0$. The local rule consists at each time step, for each cell independently, in applying the traffic rule with probability $1 - \epsilon$ and the majority rule with probability ϵ. The traffic rule (rule 184 with Wolfram's notations) is a conservative rule, which moves the 1s to the right whenever possible. It has a spacing effect. The majority rule allows the convergence to the uniform fix point, once particles have been spaced.

In order to extend this result to higher dimensions, one would like to design a rule having the same behaviour as the traffic rule, that is, to be able to compose a rule that solves the spacing problem with a majority rule. Unfortunately, this is not possible in the classical framework of cellular automata.

Proposition 2

1. In dimension 1, the traffic cellular automaton F_{184} solves the spacing problem. Precisely, it satisfies: for all $n \geq 1$ and all $x \in \{0,1\}^{\mathbb{Z}/n\mathbb{Z}}$,

$$d_q(x) \leq 1/2 \implies \forall t \geq n/2, \ F_{184}^t(x) \in \mathcal{A}_q.$$

Furthermore, $\forall q \in \{0,1\}, F_{184}(\mathcal{A}_q) \subset \mathcal{A}_q$, so F_{184} is a solution to the strict spacing problem.

2. In dimension $d \geq 2$, there are no cellular automata that solve the spacing problem.

Proof. The fact that the traffic cellular automaton spaces configurations is a "folk" result.

Let now F be a d-dimensional deterministic cellular automaton. If $x \in \mathcal{E}$ is a configuration with a symmetry of translation, then the symmetry is conserved by the evolution of the automaton. Formally, if there exists $\delta \in \mathbb{Z}^d$ such that $\forall c \in \mathcal{L}, x_c = x_{c+\delta}$, then $\forall t \in \mathbb{N}, \forall c \in \mathcal{L}, F^t(x)_c = F^t(x)_{c+\delta}$. As a consequence, deterministic cellular automata can not solve the particle spacing problem in dimension $d \geq 2$. To see why, simply consider a configuration with all 0s, except one line which is made of cells with all 1s: if the rule is conservative, this line can not disappear. □

By its very nature, a "truly" probabilistic rule can not solve the particle spacing problem. Indeed, as soon as there exists a configuration for which one cell has a non-deterministic outcome, we cannot ensure that the number of particles will be preserved. We leave open the question as to whether there exists a probabilistic rule which would solve the density classification problem with an arbitrary precision, in dimension 2 or more.

3   Particle Systems Solutions to the Spacing Problem

We have seen that deterministic cellular automata are in some sense too rigid to allow us to solve the spacing problem on grids, because they do not allow to break translation symmetries. On the other hand, probabilistic cellular automata can break these symmetries, but they do not allow an exact conservation of the number of particles.

We now propose to combine the strength of both models with interacting particle systems: we update cells by *pairs*, which allows conservation of particles, and the pairs are chosen *randomly*, which allows us to break symmetries. The effect of the local rule is to exchange the cell's states or to leave them unchanged, depending on the states of the neighbouring cells of the pair.

Let us formalize the definition of the *interacting particle systems* (IPS) we consider. From now on, we will consider two-dimensional grids. Note that most results can be adapted to higher-dimensional lattices.

3.1   Our Model of IPS

Let $\mathcal{N}_i$ and $\mathcal{N}_p$ be two finite tuples of $\mathbb{Z}^2$, corresponding to the *interaction neighbourhood* and the *perception neighbourhood*.

We define the set of interacting pairs by

$$\mathcal{I} = \{(c, c + \delta)\, ;\, c \in \mathcal{L},\ \delta \in \mathcal{N}_i\}.$$

The global rule is a function $\Phi : \mathcal{E} \times \mathcal{I} \to \mathcal{E}$. It takes in argument a configuration and a pair of cells to update, and maps it to the configuration that represents the next state of the system. The image $y = \Phi(x, (c, c'))$ is defined by:

$$(y_c, y_{c'}) = \phi\big((x_{c+k}, k \in \mathcal{N}_p), (x_{c'+k}, k \in \mathcal{N}_p)\big) \text{ and for } d \notin \{c, c'\}, y_d = x_d,$$

where $\phi : \{0,1\}^{\mathcal{N}_p} \times \{0,1\}^{\mathcal{N}_p} \to \{0,1\}^2$ is the local rule that gives the new states of the pair of cells as a function of the states of their perception neighbourhood.

This rule is conservative if the image y always satisfies $(y_c, y_{c'}) = (x_c, x_{c'})$ or $(y_c, y_{c'}) = (x_{c'}, x_c)$.

Let $(u_t)_{t \in \mathbb{N}} \in \mathcal{I}^{\mathbb{N}}$ be a sequence of interacting pairs (in the following, the u_t are chosen uniformly at random independently in $\mathcal{I}$). Starting from an initial condition $x \in \mathcal{E}$, the system will evolve according to the sequence of states (or orbit) $(x^t)_{t \geq 0}$ defined by $x^0 = x$ and $x^{t+1} = \Phi(x^t, u_t)$ for any $t \geq 0$.

Given an IPS rule Φ, we say that a set $A \subset \mathcal{E}$ is an *absorbing set* if: $\forall x \in A, \forall u \in \mathcal{I}, \Phi(x, u) \in A$. We say that A is *reachable from any configuration* if:

$$\forall x^0 \in A, \exists T \geq 0, \exists (u_t)_{1 \leq t \leq T} \in \mathcal{I}^T, x^T \in A.$$

We say that A is a *sink* if it is an absorbing set, which is reachable from any configuration.

In terms of IPS, we say that Φ is a solution to the strict spacing problem if it is a conservative IPS such that the set $\mathcal{A}$ of archipelagos is a sink.

3.2   No Solution to the Strict Spacing Problem

In order to solve the spacing problem, an idea is to design a rule, such that its evolution would result in decreasing the *energy* of the configuration, that is, the number of adjacent cells in same state. This idea will be used to propose an approximate solution in Sect. 3.4. However, the next proposition proves that this idea does not allow us to solve the strict spacing problem. This is due to the existence of configurations that are not archipelagos but for which each cell can "believe" that it is part of an archipelago (by looking at the cells located within a finite range), see Fig. 3.

Proposition 3. *There is no IPS solution to the strict spacing problem.*

The proof is omitted due to the space limitation. It can be read in the preprint version of this paper [6].

3.3   An IPS that Synchronises Checkerboards

We have seen that there is no IPS solution ensuring that once we have reached any archipelago configuration, we will remain in the set of archipelago configurations. At first sight, this could seem that there exist no solution to the spacing problem at all. However, to our surprise, we could notice that the loose problem is solvable. In fact, the *loose* problem is more demanding on the set of configurations that we do not leave. In other words, there might exist a subset $\mathcal{A}'$ of archipelagos such that once the configuration reaches $\mathcal{A}'$, it remains in it.

It can be observed that in Fig. 3, the problem comes from the fact that there are two checkerboards of different phases. If we were able to synchronize these two checkerboards, we would reach one of the two perfect checkerboards (since both 0 and 1 have density $1/2$).

For $q \in \{0, 1\}$, we denote:

$$C_e^q = \{x \in \mathcal{E};\ x_{i,j} = q \implies i + j \text{ is even}\},$$
$$C_o^q = \{x \in \mathcal{E};\ x_{i,j} = q \implies i + j \text{ is odd}\}.$$

These sets correspond to "sub-checkerboards" in the sense that in C_e^q (resp. C_o^q), state q is the minority state and only appears on even (resp. odd) cells. We also introduce $C = C_e^0 \cup C_e^1 \cup C_o^0 \cup C_o^1$. It is a subset of $\mathcal{A}$.

Proposition 4. *There is an IPS solution to the loose spacing problem for an even-sized grid. Indeed, the rule Φ_C defined below is a conservative IPS having the property that the set C is a sink.*

The proof is omitted due to the space limitation. It can be read in the preprint version of this paper [6]. We will here present the rule that solves the problem.

To define Φ_C with a local description, we introduce its interaction neighbourhood $\mathcal{N}_i = \{-1, 0, 1\}^2 \setminus \{(0,0)\}$, that is, we allow interactions between a cell and its eight nearest-neighbours. The perception neighbourhood corresponds to von Neumann neighbourhood, that is, $\mathcal{N}_p = \{(0,0), (0,1)(-1,0), (0,-1), (1,0)\}$.

Let $\mathcal{I}_4$ be the set of pairs of adjacent cells and $\mathcal{D}_4$ be the set of diagonal pairs. The interaction set is $\mathcal{I} = \mathcal{I}_4 \cup \mathcal{D}_4$. The rule will act differently on diagonal pairs on the one hand, and horizontal and vertical pairs on the other hand.

For $(i,j) \in \mathcal{I}$, let $\tau_{(i,j)} : \mathcal{E} \to \mathcal{E}$ be the function that exchanges the states x_i and x_j in configuration x. Precisely, $\tau(x, (i,j))$ is defined with:

$$\tau(x, (i,j))_k = \begin{cases} x_j & \text{if } k = i \\ x_i & \text{if } k = j \\ x_k & \text{if } k \notin \{i,j\}. \end{cases}$$

For a pair $(i,j) \in \mathcal{I}_4$, we define

$$\Phi_C(x, (i,j)) = \begin{cases} \tau(x, (i,j)) \text{ if } \text{ both cells } i \text{ and } j \text{ are not isolated,} \\ x \text{ otherwise.} \end{cases}$$

For a pair $(i,j) \in \mathcal{D}_4$, we define $\Phi_C(x, (i,j)) = \tau(x, (i,j))$.

Note that if $x_i = x_j$, both cases above result in leaving x unchanged.

In practice, numerical simulations show that in order to improve the speed of convergence to the archipelago, it is more appropriate to do the exchanges with different probability rates, depending on the state of the neighbourhood of the pair. However, these parameter do not affect the reachability properties, this is why we will here work with the simplest version of the model.

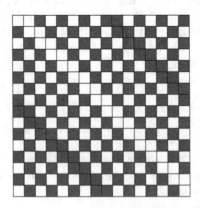

Fig. 3. Example of a configuration that locally looks like an archipelago.

3.4   Glauber Dynamics

We will now present a family of stochastic IPS having the property to solve the spacing problem "with an arbitrary precision", by converging to a distribution on configurations (with same density as the initial configuration) for which the "energy" can be controlled. Let us thus precise this notion of energy.

The energy $E(x)$ of a configuration x is the number of (horizontal or vertical) pairs 11 plus the number of (horizontal or vertical) pairs 00 in the configuration. Precisely, for $x \in \mathcal{E}$, $E(x) = \sum_{(i,j) \in \mathcal{I}_4} \mathbb{1}[x_i = x_j]$, where we recall that $\mathcal{I}_4$ is the set of pairs of adjacent cells.

Let $\mathcal{E}_k$ be the set of configurations of $\mathcal{E}$ which contain k cells in state 1.

Lemma 1. *If the grid is even-sized, then the configurations of $\mathcal{E}_k$ of minimal energy are exactly the configurations of $\mathcal{E}_k \cap \mathcal{A} = \mathcal{E}_k \cap \mathcal{A}_q$, where q is the minority state.*

The proof is omitted due to the space limitation. It can be read in the preprint version of this paper [6].

For $(i,j) \in \mathcal{I}_4$, let us also define the local energy $E_{(i,j)}(x)$ of configuration x at edge (i,j) by:

$$E_{(i,j)}(x) = \sum_{(k,\ell) \in \mathcal{V}_4(i,j)} \mathbb{1}[x_k = x_\ell],$$

where $\mathcal{V}_4(i,j)$ denotes the six edges of $\mathcal{I}_4 \setminus \{(i,j)\}$ sharing a vertex with (i,j).

Let $\beta \in \mathbb{R}$ be some fixed parameter. We propose the stochastic IPS dynamics defined as follows.

1. Choose uniformly at random a pair $u = (i,j) \in \mathcal{I}_4$ of horizontal or vertical consecutive cells.
2. Then, exchange the states of cells i and j with probability $p(x,u)$ defined by

$$p(x,u) = \frac{\exp(\beta E_u(x))}{\exp(\beta E_u(x)) + \exp(\beta(6 - E_u(x)))} = \frac{1}{1 + \exp(\beta(6 - 2E_u(x)))}.$$

The number of cells in state 1 is conserved by this dynamics, so that for any $k \in \{0, N_{\mathcal{L}}\}$, it defines a discrete time Markov chain on $\mathcal{E}_k$. The sequence of edges that will be chosen at each time step is given by a sequence of i.i.d. random variables $(u_t)_{t \in \mathbb{N}} \in \mathcal{I}_4^{\mathbb{N}}$, where u_t is uniformly distributed on $\mathcal{I}_4$. Starting from an initial condition $x \in \mathcal{E}_k$, the system evolves according to the sequence of states $(x^t)_{t \in \mathbb{N}}$ defined by $x^0 = x$ and

$$x^{t+1} = \begin{cases} \tau(x^t, u_t) & \text{with probability } p(x, u_t), \\ x^t & \text{with probability } 1 - p(x, u_t). \end{cases}$$

This Markov chain is clearly irreducible and aperiodic. We denote its transition kernel by P. In particular, if $x \neq y$ and $y = \tau(x, u)$ for some $u \in \mathcal{I}_4$, then we have: $P(x, y) = \frac{1}{\text{Card } \mathcal{I}_4} p(x, u)$.

Proposition 5. *The Markov chain defined above is reversible, and its stationary distribution on $\mathcal{E}_k$ is given by $\mu_\beta(x) = \frac{1}{Z_\beta}\exp(-\beta E(x))$ for all $x \in \mathcal{E}_k$, where $Z_\beta = \sum_{x \in \mathcal{E}_k} \exp(-\beta E(x))$.*

Proof. Let us check that the detailed balance $\mu_\beta(x)P(x,y) = \mu_\beta(y)P(y,x)$ holds for any two configurations x and y. It is enough to prove that if $x \neq y$ and $y = \tau(x,u)$ for some $u \in \mathcal{I}_4$, then $\exp(-\beta E(x))p(x,u) = \exp(-\beta E(y))p(y,u)$. But in that case, $E_u(y) = 6 - E_u(x)$ and $E(y) - E_u(y) = E(x) - E_u(x)$, so that the equality is satisfied.   □

Proposition 6. *If the grid is even-sized, then when $\beta \to \infty$, the distributions μ_β converge to the uniform measure on configurations of minimal energy, that is to the uniform measure on $\mathcal{E}_k \cap \mathcal{A}$.*

Proof. It follows from the definition of μ_β, and Lemma 1 above.   □

To sum up, the Glauber dynamics gives a simple way to approach our goal of spacing out particles. Compared to our checkerboard synchronisation rule Φ_C, it has the advantage of being simple and to use only horizontal and vertical interactions between cells. The distribution at the equilibrium can be determined analytically: it has same weight on all archipelagos with same number of particles, and the weight of non-archipelagos decreases exponentially as a function of β. But, as for Φ_C, there is still the need to know what are the time scales for observing the convergence to the equilibrium: a rule that would converge with a speed that is exponentially slow with the grid size would be useless in practice.

4   The Density Classification Problem on Finite Lattices

As a first step, we propose to study here only how to use a modified version of Φ_C to solve the density classification problem, leaving the Glauber dynamics for future work. We build our solution to the density classification problem by combining Φ_C with the majority rule.

For the checkerboard synchronisation dynamics, we introduce the following parametric variant of Φ_C, in order to increase the speed of convergence. For a configuration $x \in \mathcal{E}$ and a pair $(i,j) \in \mathcal{I}_4$, the new rule $\tilde{\Phi}_C$ is defined as follows: (a) if both i and j have exactly one adjacent cell in the same state, the exchange (which is then always allowed in Φ_C) is now applied with a probability λ; (b) in all other cases, we apply Φ_C; (c) for a pair $(i,j) \in \mathcal{D}_4$, the exchange (which is always allowed in Φ_C), is now done only with a probability χ. For $\lambda = \chi = 1$, we recover Φ_C.

We now combine $\tilde{\Phi}_C$ with a majority rule, to obtain a rule Φ_D defined as follows: for a configuration $x \in \mathcal{E}$ and a pair $(i,j) = u \in \mathcal{I}$,

$$\Phi_D(x,u) = \begin{cases} \tilde{\Phi}_C(x,u) & \text{with probability } 1-\epsilon, \\ \text{Maj}(x,i) & \text{with probability } \epsilon/2, \\ \text{Maj}(x,j) & \text{with probability } \epsilon/2, \end{cases}$$

where Maj $: \mathcal{E} \times \mathcal{L} \to \mathcal{E}$ is the function such that $y = \mathrm{Maj}(x, i)$ is defined by $y_c = x_c$ if $c \neq i$ and y_i is the majority state in the Moore neighbourhood of c (the 8 nearest neighbours of c). We can now state our main proposition.

Proposition 7. *For an even-sized grid $\mathcal{L} = (\mathbb{Z}/a\mathbb{Z}) \times (\mathbb{Z}/b\mathbb{Z})$ with $a, b \in 2\mathbb{Z}$, for any configuration $x \in \mathcal{L}$, and any non-zero value of λ and χ, the probability that Φ_{D} provides a good classification of x tends to 1 as ϵ tends to 0.*

Proof (Sketch). The proof is the same as the one given for a one-dimensional system [5]. For the sake of simplicity we can set $\lambda = 1$ and $\chi = 1$, that is, make $\tilde{\Phi}_{\mathrm{C}}$ and Φ_{C} equal. First, remark that despite the stochastic nature of our systems, archipelagos are well classified with probability one. In particular, if q is the minority state of x, then the sets of sub-checkerboards C_{e}^q and C_{o}^q are stable by the application of both $\tilde{\Phi}_{\mathrm{C}}$ and the majority rule. Moreover, the majority rule either leaves x unchanged or diminishes by 1 the number of q's in x. In other words, the system can only converge to the right fixed point $0^{\mathcal{L}}$ or $1^{\mathcal{L}}$.

The second property to remark is that as ϵ gets smaller, the probability of *not* applying the majority rule during the first k time steps tends to 1 for every value of k. In other words, for a configuration x with minority symbol q, we can make the probability to reach a sub-checkerboard in $C_{\mathrm{e}}^q \cup C_{\mathrm{o}}^q$ before applying the majority rule as high as needed. □

In order to evaluate the quality of the rule in practice, let us now briefly explore how the two rules $\tilde{\Phi}_{\mathrm{C}}$ and Φ_{D} behave with respect to their various settings.

For the particle spacing problem, our rule is defined with three parameters: the grid width L and the two probabilities of exchange λ and χ.

Let us first examine how to set λ and χ. For each setting of the system, we repeated 1000 experiments consisting of initializing the system with an independent Bernoulli of parameter $1/2$ for each cell and measuring the time needed

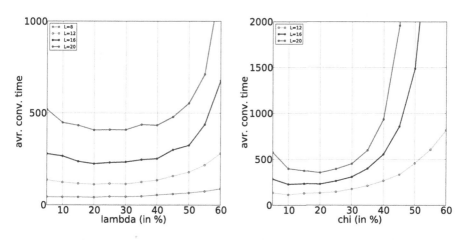

Fig. 4. Time to convergence to an archipelago for different values of L. Left: λ varies with $\chi = 0.1$. Right: χ varies with $\lambda = 0.25$

to attain an archipelago configuration. Note that for the sake of making fair comparisons, we show here the *rescaled time*, that is, a time step is taken as L^2 random updates of the global rule. We also take L to be even as for odd-sized grids there are initial conditions for which the particle spacing problem has no solution. In the previous studies, typical values of L were taken around 20, see e.g. Ref. [3,4] and references therein.

Figure 4-left shows how the average time to convergence varies as a function of λ for the specific value $\chi = 0.1$. It can be observed that for λ smaller than 0.4 the time is relatively small, while for higher values of λ, the time increases drastically.

To examine the effect of χ, we arbitrarily fixed the value of λ to 0.25 and measured the average convergence time to an archipelago. Figure 4-right shows how the average time to convergence varies as a function of χ. Here again, it can be observed that for χ smaller than 0.3 the time is relatively small, while for higher values of χ, the time increases drastically.

Interestingly, these two experiments show that in order to avoid the existence of non-archipelago fixed-point configurations, the two parameters have to be set strictly greater than zero, but can not be set too high. It is an open question to determine if there exists a phase transition with respect to the convergence to an archipelago. This would imply that, for infinite systems, if λ and χ are set above a given threshold, with a high probability, the system does not converge to an archipelago.

Our third experiment is to observe the density classification itself. For each random sample, we took random initial conditions with a uniform probability to be 0 or 1 for each cell independently. For even-sized grids, in case of equality between the number of 0s and 1s, we dismissed this initial condition and re-sampled another one with the same random distribution. We define the *quality* as the ratio of successful classifications, that is, the convergence to right fixed point $0^{\mathcal{L}}$ or $1^{\mathcal{L}}$ depending on whether the initial condition has a density smaller than or greater than $1/2$.

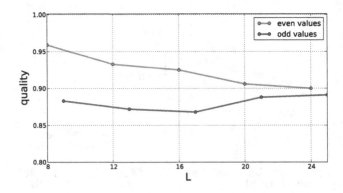

Fig. 5. Quality of classification as a function of the grid width L. (Even and odd values are shown on separate curves.) Settings are: $\lambda = 0.25$, $\chi = 0.1$, $\epsilon = 0.001$.

Figure 5 shows the evolution of the quality as a function of the grid width L, for the particular setting $\lambda = 0.25$, $\chi = 0.1$, $\epsilon = 0.001$. (Surprisingly, we empirically remarked that the quality for $\chi = 0.2$ is slightly lower.) These results show that this rule has a quality that is comparable to the best two-dimensional classification rules known so far [3,4]. Without surprise, for even-sized grids, the quality decreases as L gets larger, as it becomes more difficult to discriminate between the configurations that have approximately the same number of 0s and 1s. The curve for odd-sized grids is more surprising. Indeed, it shows an *increase* of the quality with L, at least in the range of sizes that were examined. We believe that this phenomenon results from the impossibility to space out the particles for some configurations of odd-sized grid. The system is in some sense "forced" to converge to a non-perfect configuration, in which the majority rule may introduce errors and make the system shift towards the wrong fixed point. However, as the size further increases this effect is less important and it is probable that for a given ϵ, the difference between even-sized and odd-sized grids disappears.

5   Some Questions

The density classification problem and the particle spacing problem can both be extended to infinite lattices. The set of configurations is then $\mathcal{E} = \{0,1\}^{\mathbb{Z}^d}$. For the density classification problem, a possible extension to infinite lattices consists in designing a cellular automaton on $\mathcal{E}$ such that if the initial configuration is drawn independently for each cell according to a Bernoulli law of parameter p, then if $p < 1/2$, the density of 1s converges to 0, while if $p > 1/2$, the density of 0s converges to 0.

This problem has already been studied by Marcovici and her collaborators [2]. In particular, it was shown that there is a simple example of deterministic cellular automaton that classifies the density on $\mathbb{Z}^2$: Toom's rule, which is the majority rule on the neighbourhood $\mathcal{N} = \{(0,0), (0,1), (1,0)\}$. However, in dimension 1, it is an open problem whether there exists a (deterministic or probabilistic) rule that classifies the density. Taati has partially answered this question by giving an argument that holds for densities close to zero or to one [12].

Similarly, for the particle spacing problem, an extension to infinite lattices consists in asking to design a conservative cellular automaton on $\mathcal{E}$ such that if the initial configuration is drawn independently for each cell according to a Bernoulli law of parameter p, then if $p < 1/2$, the density of non-isolated 1s converges to 0, while if $p > 1/2$, the density of non-isolated 0s converges to 0. It is known that the traffic cellular automaton F_{184} is a solution to that problem on $\mathbb{Z}$, but the problem remains open in dimension $d \geq 2$ [1].

The IPS models we have introduced in Sect. 3 are also interesting when studying them on $\mathbb{Z}^2$ instead of finite lattices. In that case, to define properly the model, we need to consider continuous-time updates: each interacting pair of $\mathcal{I}$ possess a clock that rings at times that are exponentially distributed (independently for the different pairs), and the local rule is applied when the clock rings.

It is an open problem to know if there is a proper setting of the checkerboard synchronisation dynamics having the property to space particles on $\mathbb{Z}^2$. Another interesting model is the Glauber dynamics for $\beta = \infty$. In that case, we allow exchanges between cells only if it makes the energy decrease (the exchange is made with probability $1/2$ if the exchange does not change the value of the energy). On finite grids, this IPS has many fixed points that are not archipelagos, but starting from a configuration on $\mathbb{Z}^2$ drawn according to a Bernoulli measure, the behaviour could be different.

As far as the performance of the models is concerned, we can ask what are the best settings to obtain a good trade-off between the quality of classification and the time needed to converge to a fixed point.

We also ask if we can transform our IPS into probabilistic cellular automata for solving the two-dimensional density classification problem. This can be done by using more states or by sharing the randomness of the cells (see Ref. [10] and references therein), but it is an open problem whether there is a solution within the usual framework of binary probabilistic cellular automata.

References

1. Belitsky, V., Ferrari, P.A.: Invariant measures and convergence properties for cellular automaton 184 and related processes. J. Stat. Phys. **118**(3–4), 589–623 (2005)
2. Bušić, A., Fatès, N., Mairesse, J., Marcovici, I.: Density classification on infinite lattices and trees. Electron. J. Probab. **18**(51), 1–22 (2013)
3. de Oliveira, P.P.B.: On density determination with cellular automata: results, constructions and directions. J. Cell. Automata **9**(5–6), 357–385 (2014)
4. Fatès, N.: A note on the density classification problem in two dimensions. In: Proceedings of Automata 2012, the 18th International Workshop on Cellular Automata and Discrete Complex Systems: Exploratory papers, pp. 11–18 (2012). Rapport de recherche I3S/RR-2012-04-FR - Text also available at: http://hal.inria.fr/hal-00727558
5. Fatès, N.: Stochastic cellular automata solutions to the density classification problem - when randomness helps computing. Theor. Comput. Syst. **53**(2), 223–242 (2013)
6. Fatès, N., Marcovici, I., Taati, S.: Two-dimensional traffic rules and the density classification problem (2016). https://hal.inria.fr/hal-01290290
7. Fukś, H.: Solving two-dimensional density classification problem with two probabilistic cellular automata. J. Cell. Automata **10**(1–2), 149–160 (2015)
8. Kari, J., Gloannec, B.L.: Modified traffic cellular automaton for the density classification task. Fundamenta Informaticae **116**(1–4), 141–156 (2012)
9. Land, M., Belew, R.K.: No perfect two-state cellular automata for density classification exists. Phys. Rev. Lett. **74**(25), 5148–5150 (1995)
10. Mairesse, J., Marcovici, I.: Around probabilistic cellular automata. Theor. Comput. Sci. **559**, 42–72 (2014)
11. Packard, N.H.: Adaptation Toward the Edge of Chaos. Dynamic Patterns in Complex Systems, pp. 293–301. World Scientific, Singapore (1988)
12. Taati, S.: Restricted density classification in one dimension. In: Kari, J. (ed.) AUTOMATA 2015. LNCS, vol. 9099, pp. 238–250. Springer, Heidelberg (2015)

Constant Acceleration Theorem for Extended von Neumann Neighbourhoods

Anaël Grandjean[✉]

LIRMM, Université de Montpellier, 161 rue Ada, 34392 Montpellier, France
anael.grandjean@lirmm.fr

Abstract. We study 2-dimensional cellular automata as language recognizers. We are looking for closure properties, similar to the one existing in one dimension. Some results are already known for the most used neighbourhoods, however many problems remain open concerning more general neighbourhoods. In this paper we provide a construction to prove a constant acceleration theorem for extended von Neumann neighbourhoods. We then use this theorem and some classical tools to prove the equivalence of those neighbourhoods, considering the set of languages recognizable in real time.

Introduction

Cellular automata are deterministic dynamical models. Introduced in the 1940s by S. Ulam and J. von Neumann [6] to study self replication in complex systems they were rapidly considered as computation models and language recognizers [4]. Contrary to some other classical computation models that inherently work on words, they can be considered naturally in any dimension (the original cellular automaton studied by Ulam and von Neumann were 2-dimensional) and are therefore particularly well suited to recognize picture languages. Language recognition is performed by encoding the input in an initial configuration and studying the (deterministic) evolution of the automaton from that configuration. Time and space complexities can be defined in the usual way.

One-dimensional cellular automata have been widely studied as language recognizers, and especially concerning real-time and linear time recognition. Some of the most interesting results in this field have been several closure properties as in [1,3,4]. This paper tries to expand two of those properties to 2-dimensional cellular automata.

The first theorem we present is a constant acceleration theorem for some specific set of neighbourhoods. Although such an acceleration is known in one dimension for all neighbourhoods, it was only known in two dimensions for the von Neumann and Moore neighborhoods. This result also extends the constant acceleration theorem for the von Neumann Neighbourhood, based upon a construction from V. Terrier. Although some other proofs may exists, none is currently published.

M. Cook and T. Neary (Eds.): AUTOMATA 2016, LNCS 9664, pp. 149–158, 2016.
DOI: 10.1007/978-3-319-39300-1_12

The second theorem, which is a consequence of the first one, states the equivalence, with respect to real time recognition, of some neighbourhoods (the ones for which the first theorem is true). Althought in one dimension all complete neighbourhoods are known to be equivalent [1], it has been shown by V. Terrier in [5] that at least two classes of complete neighborhoods exist in two dimensions.

1   Definitions

1.1   Cellular Automata

Definition 1 (Cellular Automaton). *A cellular automaton (CA) is a quadruple $\mathcal{A} = (d, \mathcal{Q}, \mathcal{N}, \delta)$ where*

- *$d \in \mathbb{N}$ is the dimension of the automaton;*
- *$\mathcal{Q}$ is a finite set whose elements are called states;*
- *$\mathcal{N}$ is a finite subset of $\mathbb{Z}^d$ called neighbourhood of the automaton;*
- *$\delta : \mathcal{Q}^{\mathcal{N}} \to \mathcal{Q}$ is the local transition function of the automaton.*

Definition 2 (Configuration). *A d-dimensional configuration $\mathfrak{C}$ over the set of states $\mathcal{Q}$ is a mapping from $\mathbb{Z}^d$ to $\mathcal{Q}$.*

The elements of $\mathbb{Z}^d$ will be referred to as cells and the set of all d-dimensional configurations over $\mathcal{Q}$ will be denoted as $\mathrm{Conf}_d(\mathcal{Q})$.

Given a CA $\mathcal{A} = (d, \mathcal{Q}, \mathcal{N}, \delta)$, a configuration $\mathfrak{C} \in \mathrm{Conf}_d(Q)$ and a cell $c \in \mathbb{Z}^d$, we denote by $\mathcal{N}_{\mathfrak{C}}(c)$ the neighbourhood of c in $\mathfrak{C}$:

$$\mathcal{N}_{\mathfrak{C}}(c) : \begin{cases} \mathcal{N} \to \mathcal{Q} \\ n \mapsto \mathfrak{C}(c+n) \end{cases}$$

From the local transition function δ of a CA $\mathcal{A} = (d, \mathcal{Q}, \mathcal{N}, \delta)$, we can define the *global transition function of the automaton* $\Delta : \mathrm{Conf}_d(\mathcal{Q}) \to \mathrm{Conf}_d(\mathcal{Q})$ obtained by applying the local rule on all cells:

$$\Delta(\mathfrak{C}) = \begin{cases} \mathbb{Z}^d \to \mathcal{Q} \\ c \mapsto \delta(\mathcal{N}_{\mathfrak{C}}(c)) \end{cases}$$

The action of the global transition rule makes $\mathcal{A}$ a dynamical system over the set $\mathrm{Conf}_d(\mathcal{Q})$. Because of this dynamics, in the following we will identify the CA $\mathcal{A}$ with its global rule so that $\mathcal{A}(\mathfrak{C})$ is the image of a configuration $\mathfrak{C}$ by the action of the CA $\mathcal{A}$. More generally $\mathcal{A}^t(\mathfrak{C})$ is the configuration resulting from applying t times the global rule of the automaton from the initial configuration $\mathfrak{C}$.

Definition 3 (Von Neumann and Moore Neighbourhoods). *In d dimensions, the most commonly considered neighbourhoods are the von Neumann neighbourhood $\mathcal{N}_{vN} = \{c \in \mathbb{Z}^d, ||c||_1 \leq 1\}$ and the Moore neighbourhood $\mathcal{N}_M = \{c \in \mathbb{Z}^d, ||c||_\infty \leq 1\}$. Figure 1 illustrates these two neighbourhoods in 2 dimensions.*

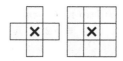

Fig. 1. The von Neumann (left) and Moore (right) neighbourhoods in 2 dimensions.

Definition 4 (a-b-Neighbourhood). *We denote by* a-b-Neighbourhood *(shortly $N_{a,b}$), the following two-dimensional neighbourhood:*

$$N_{a,b} = \{(x,y) \in \mathbb{Z}^2 \mid b|x| + a|y| \leq ab\}$$

Note that such a neighbourhood is convex and symmetric with respect to the origin. For completeness reasons we also require a and b to be strictly positive. Furthermore, the neighbourhood $N_{1,1}$ is exactly the von Neumann Neighbourhood. Some examples are depicted in Fig. 2.

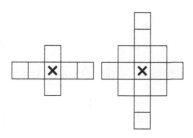

Fig. 2. $N_{2,1}$ (left) and $N_{2,3}$ (right)

1.2 Picture Recognition

From now on we will only consider 2-dimensional cellular automata (2DCA), and the set of cells will always be $\mathbb{Z}^2$.

Definition 5 (Picture). *For $n, m \in \mathbb{N}$ and Σ a finite alphabet, an (n,m)-picture (picture of width n and height m) over Σ is a mapping*

$$p : [\![0, n-1]\!] \times [\![0, m-1]\!] \to \Sigma$$

$\Sigma^{n,m}$ *denotes the set of all (n,m)-pictures over Σ and $\Sigma^{*,*} = \bigcup_{n,m \in \mathbb{N}} \Sigma^{n,m}$ the set of all pictures over Σ. A picture language over Σ is a set of pictures over Σ.*

Definition 6 (Picture Configuration). *Given an (n,m)-picture p over Σ, we define the picture configuration associated to p with quiescent state $q_0 \notin \Sigma$ as*

$$\mathfrak{C}_{p,q_0} : \begin{cases} \mathbb{Z}^2 \to \Sigma \cup \{q_0\} \\ x, y \mapsto \begin{cases} p(x,y) & if(x,y) \in [\![0, n-1]\!] \times [\![0, m-1]\!] \\ q_0 & otherwise \end{cases} \end{cases}$$

Definition 7 (Picture Recognizer). *Given a picture language L over an alphabet Σ, we say that a 2DCA $\mathcal{A} = (2, \mathcal{Q}, \mathcal{N}, \delta)$ such that $\Sigma \subseteq \mathcal{Q}$ recognizes L with quiescent state $q_0 \in \mathcal{Q} \setminus \Sigma$ and accepting states $\mathcal{Q}_a \subseteq \mathcal{Q}$ in time $\tau : \mathbb{N}^2 \to \mathbb{N}$ if, for any picture p (of size $n \times m$), starting from the picture configuration $\mathfrak{C}_{p,q_0}$ at time 0, the origin cell of the automaton is in an accepting state at time $\tau(n, m)$ if and only if $p \in L$. Formally,*

$$\forall n, m \in \mathbb{N}, \forall p \in \Sigma^{n,m}, \quad \mathcal{A}^{\tau(n,m)}(\mathfrak{C}_{p,q_0})(0,0) \in \mathcal{Q}_a \Leftrightarrow p \in L$$

We then say that the language L can be recognized in time $\tau(n, m)$ with neighbourhood N.

Since cellular automata work with a finite neighbourhood, the state of the origin cell at time t (after t actions of the global rule) only depends on the initial states on the cells in $\mathcal{N}^t$, where $\mathcal{N}^0 = \{0\}$ and for all n, $\mathcal{N}^{n+1} = \{x + y, \, x \in \mathcal{N}^n, y \in \mathcal{N}\}$. The real time function is informally defined as the smallest time such that the state of the origin may depend on all letters of the input:

Definition 8 (Real Time). *Given a neighbourhood $\mathcal{N} \subset \mathbb{Z}^d$ in d dimensions, the real time function $\tau_\mathcal{N} : \mathbb{N}^d \to \mathbb{N}$ associated to $\mathcal{N}$ is defined as*

$$\tau_\mathcal{N}(n_1, n_2, \ldots, n_d) = \min\{t, [\![0, n_1 - 1]\!] \times [\![0, n_2 - 1]\!] \times \ldots \times [\![0, n_d - 1]\!] \subseteq \mathcal{N}^t\}$$

When considering the specific case of the 2-dimensional von Neumann neighbourhood, the real time is defined by $\tau_{\mathcal{N}_{vN}}(n, m) = n + m - 2$. There is however a well known constant speed-up result:

Proposition 1 (folklore). *For any $k \in \mathbb{N}$, any language that can be recognized in time $(\tau_{\mathcal{N}_{vN}} + k)$ by a 2DCA working on the von Neumann neighbourhood can also be recognized in real time by a 2DCA working on the von Neumann neighbourhood.*

So, it will be enough to prove that a language is recognized in time $(n, m) \mapsto n + m + k$ for some constant k to prove that it is recognized in real time.

2   Main Result

Theorem 1 (constant acceleration). *For any a, b and k positive integers, any language that can be recognized in time $(\tau_{N_{a,b}} + k)$ by a 2DCA working on $N_{a,b}$ can also be recognized in real time by a 2DCA working on the same neighbourhood.*

Proof. To prove the theorem one only needs to be able to accelerate one step of the calculation (that is, recognize in real time any language recognized in real time plus one). We then only have to repeat the process a finite number of times.

As in the one dimensional case, the idea is to make each cell "guess" the state of some further cells, so that at real time, the origin cell guesses are correct, and allow it to perform one more step of computation.

Fix some language L and some automaton A recognizing this language in real time plus one step. This automaton works on the a-b-Neighbourhood $N_{a,b}$. We will now construct an automaton A' with the same neighbourhood, recognizing L in real time.

The potential initial configurations of both automata are the same, from now on we fix one initial configuration and explain what happens on the run starting from this configuration in A', depending on what happens in the run starting from this same configuration in A.

Let us introduce some notations: $A(c,t)$ represents the state of the cell c at time t in the original automaton A. We denote as G_{all} the following set:

$$G_{all} = \{(xa, yb) \in \mathbb{Z}^2 \| 1 < x + y \leq 2; 0 \leq x; 0 \leq y; x \leq y + 1; y \leq x + 1\}$$

This is a subset of the north east quarter of N^2, depicted in green in Fig. 3. Similarly we define G_{left} and G_{bottom} as follows, depicted in light blue in Fig. 3:

$$G_{left} = \{(xa, yb) \in \mathbb{Z}^2 \| 1 < x + y \leq 2; 0 \leq x; 0 \leq y; x \leq y + 1; y \geq x + 1\}$$

$$G_{bottom} = \{(xa, yb) \in \mathbb{Z}^2 \| 1 < x + y \leq 2; 0 \leq x; 0 \leq y; x \geq y + 1; y \leq x + 1\}$$

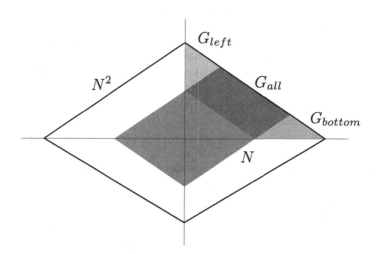

Fig. 3. A partition of N^2 (Color figure online)

Each cell c of the new automaton A' will "contain" many states of cells of A. More formally the state set of A' is a power of the state set of A. A cell c of A' contains, at time $t \geq 1$, the following informations:

- $A(c,t)$
- $A(c', t-1)$ for every $c' \in N(c)$.
- $g(c, c', t-1)$ for every $c' \in G_{all}(c)$.

- $g(c, c', t-1)$ for every $c' \in G_{bottom}(c)$ if c is on the bottom border of the input word.
- $g(c, c', t-1)$ for every $c' \in G_{left}(c)$ if c is on the left border of the imput word.

We call $E(c)$ (extended neighbourhood) the set of cells c' such that c holds either $A(c', t)$ or $g(c, c', t)$. Remark that $E(0)$ is exactly the north east quarter of N^2.

The state $g(c, c', t)$ is some sort of guess, made by the cell c, of what could be $A(c', t)$. For every cells c and c', $g(c, c', 0) = \#$, the quescient state of A. The update rule of $g(c)$ will depend on the available information for the cell c. The new state $g(c, c', t+1)$, is the result of applying the local rule of A on the neighbours of c', using the state of A when it is available to c, and a guess of another cell otherwise.

To better understand this update rule, we need to focus on what information is available for each cell of A' during the computation. A cell c can "see" every information contained its neighbouring cells. This way, for example, at time t, a cell c have access to all $A(c', t-1)$ for $c' \in N^2(c)$, as each cell of its neighbourhood contain this information for each cell of its own neighbourhood. This way it is easy to update every state of A, directly using the local rule of A.

Remark that one cell can see several times the same information, as it is contained in more than one cell of its neighbourhood. As the information about the states of $A(c, t)$ results from a direct simulation, all its occurences are coherent. On the contrary, guesses about one cell state can depend on which cell is making the guess. To avoid incoherence issues, a cell c will only use the information contained in the guesses of its leftmost and uppermost neighbours, and completely ignore all other guesses (including its own previous guesses).

Some quick math calculations shows that the neighbourhood of each cell in $E(c)$ is indeed contained in $N^2(c) \cup E(c + (a, 0)) \cup E(c + (0, b))$, as depicted in Fig. 4. A cell whose state is $\#$ formally stays in this state, but is "read" as if its extended neighbourhood was filled with $\#$.

This equation is not true for the cells on the left border and on the bottom border, those cells needs to use some guesses about cells outside the computation. However those cells will always remain in the quescient state, therefore no information is actually missing.

We will say that a cell is correct if all the $g(c, c', t)$ in its extended neighbourhood are equal to $A(c', t)$. After the first time step, all the cells of the upmost row and of the rightmost column are correct. Moreover all the cells with the top right corner cell in its neighbourhood are correct. As all the cells outside the computation also correct, we can note that each cell which is above or on the right of a correct cell is also correct.

At time $t+1$ a cell c is correct if both $c + a$ and $c + b$ are correct. This is also true for the cells on the borders, using the fact that every cell outside the border is in state $\#$.

Because of the shape of the set of correct cells at time 1, it is easy to see that a time t, the set of correct cells will contain $N^{-t}((n, m))$ where n and m are respectively the length and height of the input.

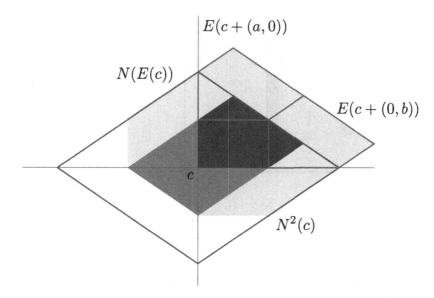

Fig. 4. Available and needed information for updating cells

Because of the shape of the neighbourhood, this cell is the farthest from the origin. Then, the real time is equal to the minimal t such that the origin cell is in $N^{-t}(n, m)$. Thus, at real time, denoted $\tau(n, m)$, the origin cell is correct, and therefore knows the state of every cell of its extended neighbourhood at time $\tau - 1$, in the automaton A. The extended neighbourhood of the origin consists of all the cells in $N^2(0)$ which are not in state $\#$. This information is enough to compute the state of all cells in $N((0, 0))$ at time τ in A. Then it is engouh information to compute the state of the origin cell at time $\tau + 1$ in A.

The accepting states of A' are all the "configurations" of the extended neighbourhood of the origin cell which lead to an acceptation in A at time $\tau + 1$.

Theorem 2 (Equivalence of the a-b-Neighbourhoods). *For any a, b, c, d positive integers, any language that can be recognized in real time by an automaton with $N_{a,b}$ can also be recognized in real time by an automaton with $N_{c,d}$.*

Proof. The proof of this theorem is based upon the two following lemmas:

Lemma 1. *For any a, b and k positive integers, the sets of languages recognized in real time with $N_{a,b}$ and with $N_{a,b}^k$ are the same. We say that $N_{a,b}$ and $N_{a,b}^k$ are* equivalent *with respect to the real time.*

Lemma 2. *For any a, b and k positive integers, any language recognized in real time with $N_{ka,b}$ (symmetrically $N_{a,kb}$) can also be recognized in real time with $N_{a,b}$. We say that $N_{a,b}$ is* more powerful *than $N_{ka,b}$.*

The first lemma is well known for one dimensional cellular automata, and is an easy corollary of the constant acceleration theorem. First note that the real time function for N and the one for N^k are very similar. Indeed $\tau_{N^k} = \lceil \frac{\tau_N}{k} \rceil$.

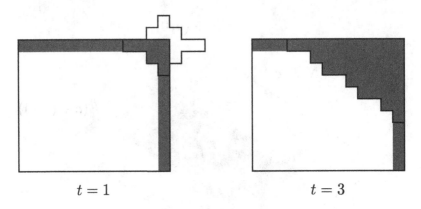

$$t = 1 \qquad\qquad\qquad t = 3$$

Fig. 5. Correct cells at the first steps

As an automaton A' with neighbourhood N^k can simulate k steps of computation of an automaton A with neighbourhood N in a single step, in real time it can simulate at least τ_N steps of A, proving one inclusion (Fig. 5).

The other inclusion needs the constant acceleration theorem. Indeed, an automaton A' with neighbourhood N needs k steps to simulate one step of an automaton A with the neighbourhood N^k. Thus it needs at most $\tau_N + k$ steps to compute τ_{n^k} steps of A. Thanks to the first theorem of this paper, we can build another automaton with neighbourhood N recognizing the same language in real time, completing the proof.

In order to prove the second lemma, we will have to perform a compression of the input. Let A be an automaton recognizing L with neighbourhood $N_{ka,b}$ in real time. Once again the two neighbourhoods we consider have very similar real time:

$$\tau_{N_{ka,b}}(n, m) = \lceil \frac{n}{ka} \rceil + \lceil \frac{m}{b} \rceil$$

$$\tau_{N_{a,b}}(n, m) = \lceil \frac{n}{a} \rceil + \lceil \frac{m}{b} \rceil$$

Now consider the following neighbourhood:

$$M = \{(x,0)| -ka \le x \le ka\} \cup \{(0,y)| -b \le y \le b\}$$

M have exactly the same convex hull as $N_{ka,b}$, but is not convex. However thanks to Delacourt and Poupet [2] we know that there is an automaton A' with neighbourhood M which recognizes L in time at most $\tau_{N_{ka,b}} + c$ for some constant c.

We will now build an automaton B which simulates A'. Each cell of B will be able to store up to k states of A'. First the automaton will perform a compression of the input, line by line, by a factor k. This takes $\lceil \frac{((k-1)n)}{ka} \rceil$ time steps.

After the compression each cell of B contains k cells of A', and can simulate the computation of A' for each of those cells without losing any time. In Fig. 6 the automaton B is represented after a compression by factor 3. Each blue

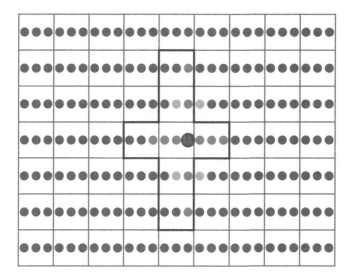

Fig. 6. Available information after compression (Color figure online)

dot in a cell correspond to a cell of A'. Here, $N_{1,2}$ is the neighbourhood of B, depicted in red. $N_{3,2}$ is the neighbourhood of A. The green dots represents the neighbourhood of the blue dot circled in green. The four light green dots are the one which are in $N_{3,2}$ but not in M with respect to the circled dot.

By doing this simulation, at time $\tau_{N_{a,b}} + c$ the automaton B have simulated $\tau_M + c$ steps of A', recognizing language L. Thanks to the constant acceleration theorem, there is an automaton B' which can also recognize L in real time with neighbourhood $N_{a,b}$.

Now lets go back to the proof of our theorem. Consider four integers a, b, c and d, N the a-b-Neighbourhood, and M the c-d-Neighbourhood. We denote by M' the abc-abd-Neighbourhood and N' the abc-b-neighbourhood. By the second lemma, we know that $N_{a,b}$ is more powerful than $N_{abc,b}$. By applying this lemma again we have that $N_{abc,b}$ is more powerful than $N_{abc,abd}$. Therefore $N_{a,b}$ is more powerful than $N_{abc,abd}$. Remark that $N_{c,d}^{ab} = N_{abc,abd}$. Thanks to the first lemma we know that $N_{c,d}$ and $N_{abc,abd}$ are equivalent with respect to real time recognition. Thus $N_{a,b}$ is more powerful than $N_{c,d}$.

With similar ideas we can prove that $N_{c,d}$ is more powerful than $N_{a,b}$, proving the announced result.

References

1. Poupet, V.: Cellular automata: real-time equivalence between one-dimensional neighborhoods. Theor. Comput. Syst. **40**(4), 409–421 (2007)
2. Delacourt, M., Poupet, V.: Real time language recognition on 2D cellular automata: dealing with non-convex neighbourhoods. In: Kučera, L., Kučera, A. (eds.) MFCS 2007. LNCS, vol. 4708, pp. 298–309. Springer, Heidelberg (2007)

3. Mazoyer, J., Reimen, N.: A linear speed-up theorem for cellular automata. Theor. Comput. Sci. **101**, 59–98 (1991)
4. Smith III, A.R.: Real-time language recognition by one-dimensional cellular automata. J. ACM **6**, 233–253 (1972)
5. Terrier, V.: Two-dimensional cellular automata recognizer. Theor. Comput. Sci. **218**(2), 325–346 (1999)
6. von Neumann, J.: Theory of Self-Reproducing Automata. University of Illinois Press, Urbana (1966)

Shrinking and Expanding Cellular Automata

Augusto Modanese and Thomas Worsch[(✉)]

Institute for Theoretical Informatics, Karlsruhe Institute of Technology,
Am Fasanengarten 5, 76131 Karlsruhe, Germany
augusto.modanese@student.kit.edu, worsch@kit.edu

Abstract. Inspired by shrinking cellular automata (SCA), we investigate another variant of the classical one-dimensional cellular automaton: the shrinking and expanding cellular automaton (SXCA). In addition to the capability to delete some cells as in SCA, an SXCA can also create new cells between already existing ones. It is shown that there are reasonably close (polynomial) relations between the time complexity of SXCA and the space and time complexity of Turing machines and alternating Turing machines respectively. As a consequence the class of problems decidable in polynomial time by SXCA coincides with **PSPACE**.

1 Introduction

Rosenfeld et al. [5] have introduced so-called shrinking cellular automata.

The idea is that the transition function allows a cell to "delete" itself. If one cell does that, then its both neighbors to the left and right become direct neighbors, and similarly if several cells are deleted (a precise definition will be given in Sect. 2). Rosenfeld and Wu observed that some formal languages can be recognized in less than real-time, using the shrinking process to mimic reductions according to the productions of a context free grammar.

It took more than 30 years before Kutrib et al. [3] started a more thorough investigation of shrinking CA, explaining relations between shrinking and nonshrinking CA for small time bounds, i. e. real-time and linear time.

It seemed natural to also have a first look at cellular automata which can not only shrink, but also have the opposite ability expand by generating additional cells between already existing ones. As in the shrinking case, at least at first sight, it is only clear how to define this in the one-dimensional case. It turns out that shrinking and expanding CA are quite powerful.

The remainder of this paper is organized as follows. In Sect. 2 we define shrinking and expanding cellular automata and related notions like the set **SXCAP** of problems decidable by SXCA in polynomial time. In Sect. 3 we relate time complexity of alternating Turing machines to that of SXCA, implying that **PSPACE** is included in **SXCAP**. In Sect. 4 we obtain results showing that in particular **SXCAP** in included in **PSPACE**. (Hence SXCA belong to the so-called second machine class [6].)

This paper is partially based on the bachelor thesis by the first author [4].

© IFIP International Federation for Information Processing 2016
Published by Springer International Publishing Switzerland 2016. All Rights Reserved
M. Cook and T. Neary (Eds.): AUTOMATA 2016, LNCS 9664, pp. 159–169, 2016.
DOI: 10.1007/978-3-319-39300-1_13

2   Basics

The set of all integers is denoted $\mathbb{Z}$, the set of all positive integers $\mathbb{N}_+$ and $\mathbb{N}_0 = \mathbb{N}_+ \cup \{0\}$. For two sets A and B the set of all total functions mapping from A to B is denoted B^A.

2.1   Standard CA

In this paper we are looking at one-dimensional cellular automata with the standard neighborhood $N = \{-1, 0, 1\}$ of radius 1. Let Q denote the set of states of a single cell.

Then a global configuration is a function $c \colon \mathbb{Z} \to Q$. The local transition function $\delta \colon Q^N \to Q$ induces a global transition function $\Delta \colon Q^{\mathbb{Z}} \to Q^{\mathbb{Z}}$ in the usual way: $\Delta(c)(x) = \delta(\ell_{c,x})$ where $\ell_{c,x} \colon N \to Q \colon n \mapsto c(x+n)$.

A (finite or infinite) *computation* is a sequence $(c_0, c_1, \dots)$ of configurations such that $c_{t+1} = \Delta(c_t)$ holds for all t.

We will always assume that there is a special state q such that the local transition functions satisfies two properties:

- As long as the cells in the neighborhood of a cell x are all in state q the next state of x is q again, i.e. q is a quiescent state.
- A cell which is not in state q will never enter state q in the next step.

We will only look at decision problems. The input for a CA is always a non-empty word $w \in A^+$ over some finite alphabet $A \subset Q$. The initial configuration of a word $w = a_0 \cdots a_{n-1}$ of length $|w| = n$ with $a_i \in A$ for all $i \leq |w|$ is defined as

$$c_w \colon \mathbb{Z} \to Q \colon x \mapsto \begin{cases} a_i, & \text{iff } 0 \leq x = i < |w| \\ q, & \text{otherwise} \end{cases}. \tag{1}$$

The requirements for q ensure that all configurations reachable from initial configurations c_w as just defined all consist of a finite connected block of cells all of which are in a non-quiescent state, extended on both sides with infinitely many cells in state q. Of course the block of non-quiescent cells may grow during a computation.

For the acceptance and rejection of input words we assume that a special state a ("accept") and a special state r ("reject") are present in Q. A configuration c is *final* iff cell 0 is in the accepting or the rejecting state. The computation $(c_w = c_0, c_1, \dots, c_t)$ for an input w is halting if c_t is final and if c_t is the only final configuration in the computation. We call c_t the result configuration for w. An input is accepted or rejected iff in the result configuration cell 0 is accepting or rejecting respectively.

The *support* of a configuration c is the set $\{ x \mid c(x) \neq q \}$. Let $\mathcal{C}_{fin}$ denote the set of all configurations $c \colon \mathbb{Z} \to Q$ with finite support.

2.2   Shrinking and Expanding CA

We now have a look at non-standard CA which can *shrink* by deleting cells "not needed" any longer [3,5] and which can *expand* by generating new cells between already existing ones [4].

In the present paper the focus is on CA which are allowed to do both during a computation. The following definition of *shrinking and expanding CA (SXCA)* is adapted to this case.

Besides the set Q of proper states there is a "pseudo state" $\otimes$. Formally we assume $\otimes \notin Q$ and write $\bar{Q} = Q \cup \{\otimes\}$. In analogy to $\mathcal{C}_{fin}$ let $\bar{\mathcal{C}}_{fin}$ denote the set of all functions $\bar{c}: \mathbb{Z} \to \bar{Q}$ with having finite support $\{x \mid \bar{c}(x) \neq q\}$.

The local transition function $\delta = (\varepsilon, \sigma)$ of an SXCA is given by two functions $\varepsilon: Q^N \to \bar{Q}$ and $\sigma: Q^N \to \bar{Q}$. Given a configuration $c \in \mathcal{C}_{fin}$ these induce a global transition of the SXCA defined in two substeps.

– To a given configuration c first ε and σ are applied to each local configuration $\ell_{c,x}$ observed by cell x in c. This gives rise to a function $X: \mathcal{C}_{fin} \to \bar{\mathcal{C}}_{fin}$ in the following way.

$$\text{for all } x \in \mathbb{Z}: \quad \begin{aligned} X(c)(2x-1) &= \varepsilon(\ell_{c,x}) \\ X(c)(2x) &= \sigma(\ell_{c,x}) \end{aligned} \quad (2)$$

One can imagine that the existing cells are pulled apart by doubling their indices entering the preliminary states $\sigma(\ell_{c,x})$ respectively. To the left of each such cell a "new" cell is generated with preliminary state $\varepsilon(\ell_{c,x})$.

– Secondly, the cells which are in state $\otimes$ "are removed" and the "remaining cells are renumbered" in the following sense: Given any $\bar{c} \in \bar{\mathcal{C}}_{fin}$, define a renumbering

$$\begin{aligned} r_{\bar{c}}&: \mathbb{Z} \to \mathbb{Z} \\ 0 &\mapsto \min\{x' \geq 0 \mid \bar{c}(x') \in Q\} \\ x &\mapsto \min\{x' > r_{\bar{c}}(x-1) \mid \bar{c}(x') \in Q\}, \quad \text{if } x > 0 \\ x &\mapsto \max\{x' < r_{\bar{c}}(x+1) \mid \bar{c}(x') \in Q\}, \quad \text{if } x < 0 \end{aligned} \quad (3)$$

This induces a function $S: \bar{\mathcal{C}}_{fin} \to \mathcal{C}_{fin}$ by

$$S(\bar{c})(x) = \bar{c}(r_{\bar{c}}(x)) \quad (4)$$

Figure 1 shows an example.

The composition $\Delta = S \circ X$ is the global transition function of the SXCA.

CA in the classical definition are a special case, where existing cells are never deleted, i.e. all $\sigma(\ell) \neq \otimes$, and additional cells are never generated, i.e. all $\varepsilon(\ell) = \otimes$. Shrinking CA are recovered by again requiring all $\varepsilon(\ell) = \otimes$ and interpreting $\otimes$ as the "dissolve" value used in [3].

The definitions for the notions *computation, acceptance*, etc., are defined in the same way as for standard CA.

The *time complexity* of an SXCA is the function $f: \mathbb{N}_+ \to \mathbb{N}_+$ where $f(n)$ is the maximum number of steps needed for an input of length n until it is accepted or rejected. The set of problems which can be decided by SXCA with polynomial time complexity will be written **SXCAP**.

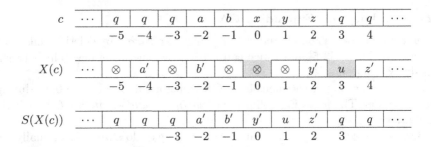

Fig. 1. The two substeps of a global transition of an SXCA: The cells in states a, b and y should enter states $a' = \sigma(q, a, b)$, $b' = \cdots$ and $y' = \cdots$ respectively without generating an additional cell to their left ($\varepsilon(q, a, b) = \otimes, \ldots$). The cell in state x deletes itself (left gray cell, $\sigma(b, x, y) = \otimes$) and also doesn't generate an additional cell to its left ($\varepsilon(b, x, y) = \otimes$). The cell in state z enters $z' = \sigma(y, z, q)$ and generates a cell in state $u = \varepsilon(y, z, q)$ to its left (second gray cell). Cells which are "not needed" are deleted and the remaining ones "renumbered" during the second substep.

3   Relating ATM Time to SXCA Time

Fellah and Yu [2] have shown an intricate close relation between tree-shaped cellular automata (with sequential input at the root) and alternating Turing machines (ATM) [1].

While the repeated generation of additional cells in an SXCA can yield an exponential growth of the number of cells —which superficially looks similar to the exponentially growing number of cells reachable from the root of a tree— there is a difference: In a tree the root can quickly broadcast a value in t steps to 2^t cell. On the other hand, in a growing CA a cell (the "root") from which the generation of 2^t cells has been initiated, *cannot* in t steps broadcast a value to 2^t cells (once the generation process is complete). This seems to increase the complexity of some constructions. Still, we will prove in this section:

Theorem 1. *An alternating TM with time complexity bounded by some function $t(n)$ can be simulated by an SXCA whose time complexity is $O((t(n))^2)$.*

Since alternating time is polynomially related to sequential space [1], one immediately obtains:

Corollary 2. PSPACE $\subseteq$ SXCAP, *i.e. the set of problems which can be decided by SXCA in polynomial time comprises* **PSPACE**.

We now describe a construction proving Theorem 1. We assume the reader is familiar with ATMs.

Let T be an alternating Turing machine with one work tape on which also the input is provided at the beginning of a computation. Furthermore let S denote the set of states, B the set of tape symbols, and $A \subset B$ the input alphabet of the ATM. One may assume that in each situation there are exactly two possible

[			s				]
[	a	b	c	d	e	f	]

Fig. 2. A block storing 6 tape symbols. The head of the ATM is visiting the third square and the ATM is in state s.

[	[			s				]	∃/∀	[			s'				]	]
[	[	a	b	c	d	e	f	]		[	a	b	c	d	e	f	]	]

Fig. 3. Two copies of a block storing 6 tape symbols. The head of the ATM is visiting the third square and the ATM is in state s. Two "versions" of state s are used in order to allow the simulation of the ATM for its two possible continuations. The extra cell between the block remembers whether s is an existential or a universal state.

continuations for the ATM. We will construct an SXCA C which for each input $w \in A^+$ checks whether T would accept or reject w and then accepts or rejects w accordingly.

Without loss of generality T never writes blank symbols, and hence there is always one contiguous tape segment with non-blank symbols comprising all (input and) visited squares. The SXCA C uses k consecutive cells to store such segment of length k with symbols $a_1 \cdots a_k$, along with a representation of the ATM head being positioned on symbol a_p, and the ATM being in state $s \in S$. Each cell stores one symbol, and cell p in addition the state s while the others store a value indicating that the head is not on the respective square. In addition the block is surrounded by two "bracket cells". See Fig. 2 for an example of such a *block* of the SXCA.

In its very first step, C changes the states storing the input symbols such that the cells represent the corresponding initial ATM configuration (as described above). Afterwards, C iterates the following procedure:

1. New cells are generated to the left of the block and the block contents are copied to it. For each state s of the ATM, the states in both blocks are distinguished by using an unprimed representation s in the left one, and a primed representation s' in the right one.
2. Between the blocks an additional cell is used to remember whether the ATM state is an existential or a universal one. The pair of blocks is surrounded by an additional pair of brackets. See Fig. 3.
3. In the left block, the first alternative for the next step of the ATM is simulated, while the second alternative is simulated in the right one. The SXCA can distinguish between the two cases because different representations of the state are used in the two blocks.
4. In each block C checks whether the ATM state is a final one.
   - If this is not the case, then C continues with step 1 above.
   - If the state is final, C breaks out of the loop for the block and continues with step 5 below.

5. If the ATM state is a final one, then signals are sent in both directions to the nearest enclosing brackets deleting all cells which do not store the state or the brackets. Thus, the block is shrunk to a single cell storing state a or state r.

6. Eventually, an extra cell storing a quantifier will observe final states in both its neighboring cells. It then sends signals in both directions to the nearest enclosing brackets. This causes the deletion of the two cells which contain final states, as well as of the ones containing the brackets. The cell itself enters state a or r according to the following rules:

   – It enters a in case the quantifier is $\exists$ and at least one of the neighboring final states was a, or if the quantifier is $\forall$ and both neighboring final states were a.
   – Otherwise the cell enters state r.

As long as a configuration is not a final one, the block, inside the brackets, representing it will be replaced by two bracketed blocks and the $\exists/\forall$ marker. The resulting nested hierarchy of brackets will be deleted from the inside out, once final configurations have been reached.

In the end, all non-quiescent cells are deleted except for a single cell, which is in state a or r if and only if the ATM would accept or reject the input, respectively.

It remains to estimate the time complexity of the SXCA. If the ATM is $t(n)$ time-bounded, then, for an input of length n, it will never make more than $t(n)$ steps during a possible computation and hence never visit more than $t(n)$ squares. Therefore, the length of all blocks which have to be copied during step 1 is at most $t(n)$. The simulation of one step of the ATM requires only constant time. Consequently after $O((t(n))^2)$ SXCA steps in every block a final state is reached.

Deleting the non-state cells within a block again only takes time $t(n)$, and this also has to be done $t(n)$ times. Therefore the result of the ATM is available after $O((t(n))^2)$ steps.

4   Relating SXCA Time to TM Space

At the beginning of the previous section it has already been pointed out that the (compared to tree CA) somewhat restricted possibilities of communication between cells may manifest themselves in somewhat less efficient simulations of other models by SXCA. On the other hand, the ability to delete a possibly large number of cells simultaneously, until now seems to impede particularly efficient "simulations" of SXCA by other models such as ATM. Instead of elaborating a somewhat arduous ATM, in this section we only exhibit a space efficient approach:

Theorem 3. *An SXCA whose time complexity is bounded by $t(n)$ can be simulated by a deterministic TM with space complexity bounded by $O((t(n))^2)$.*

Together with Theorem 1 this implies:

Corollary 4. SXCAP = PSPACE, *i. e. the set of problems which can be decided by SXCA in polynomial time is exactly* **PSPACE.**

Thus SXCA are a model in the so-called *second machine class* [6].

In order to prove Theorem 3 we first describe an algorithm and discuss its space complexity afterwards. For better readability at some points we have forced page breaks to make sure that pseudo code and corresponding explanations are on the same page.

We proceed "top down", starting with a very simple function, and proceed in three refining steps.

Assuming that there is a function STATE(cell x, time t) which computes the state of cell $x \in \mathbb{Z}$ at time $t \in \mathbb{N}_0$, it is trivial to determine whether the SXCA accepts a given input w. One just has to find the first time, when cell 0 enters a final state and check that. See Algorithm 1.

⟨ FINALSTATE returns a or r if the SXCA accepts or rejects an input w, respectively. ⟩
function *final state* ← FINALSTATE() **is**
 $\quad\mid\quad t \leftarrow 0$
 $\quad\mid\quad$ **repeat**
 $\quad\mid\quad\mid\quad t \leftarrow t + 1$
 $\quad\mid\quad\mid\quad s \leftarrow \text{STATE}(0, t)$
 $\quad\mid\quad$ **until** $s \in \{a, r\}$
 $\quad\mid\quad$ **return** s
end

Algorithm 1. Finding the final result of an SXCA.

Next, we describe the process of determining the state of a cell x at time t. The case $t = 0$ is trivial. For $t > 0$, it is important to remember how the transition from a configuration c to $S(X(c))$ was defined in Sect. 2; in particular see Fig. 1. There are two possibilities how a cell z at time $t - 1$, given its state and those of its both neighbors, can give rise to the state of cell x at time t: (i) by changing its own state using σ; or (ii) by generating a new cell using ε. Assume that z, along with a flag *new* which signals whether x was created using ε or not, can be computed by a function PREDECESSOR(x, t). Then Algorithm 2 shows a straightforward way to compute the state of cell x at time t.

The renumbering of cells was defined earlier by distinguishing between two cell groups: the ones with indices $x \geq 0$, and those with indices $x < 0$. Observe that the renumbering of cells never changes the index of a cell from one group to the other one. Now, consider the case $x \geq 0$ first, and assume that there is function SUCCESSOR(x, t) which, for a cell index x at time t, returns the index z at which the given cell will be located at time $t + 1$, or ∞ in case the cell is deleted. Then, determining PREDECESSOR(x, t) is a simple matter of counting how many cells z get positioned to the left of x at time t. The case $x < 0$ is similar; the difference is that one has to count the cells to the right. Using the

⟨ Compute the state of cell x at time t: ⟩
function $state\ s \leftarrow \text{STATE}(cell\ x, time\ t)$ **is**

> **if** $t = 0$ **then**
>> ⟨ Initial configuration for input w: ⟩
>> **if** $0 \le x < |w|$ **then**
>>> | **return** w_x ⟨ the x-th symbol of input w ⟩
>>
>> **else**
>>> | **return** q
>>
>> **end**
>
> **else**
>> $(z, new) \leftarrow \text{PREDECESSOR}(x, t)$
>> $q_{-1} \leftarrow \text{STATE}(z - 1, t - 1)$
>> $q_0 \leftarrow \text{STATE}(z, t - 1)$
>> $q_1 \leftarrow \text{STATE}(z + 1, t - 1)$
>> **if** new **then**
>>> ⟨ cell was "generated" during transition to time t ⟩
>>> **return** $\varepsilon(q_{-1}, q_0, q_1)$
>>
>> **else**
>>> ⟨ cell already existed ⟩
>>> **return** $\sigma(q_{-1}, q_0, q_1)$
>>
>> **end**
>
> **end**

end

Algorithm 2. Determining the state of a cell x at time t.

⟨ Compute cell index z such that the states of cells $(z - 1, z, z + 1)$ at time $t - 1$ determine the state of cell x at time t; additionally return a flag new indicating whether the cell now at position x is "a new one" or not, i.e. whether ε of σ should be used to compute its state. ⟩
function $(cell\ z,\ bool\ new) \leftarrow \text{PREDECESSOR}(cell\ x, time\ t)$ **is**

> $z \leftarrow 0$
> **while** $true$ **do**
>> **if** $\text{SUCCESSOR}(z, t - 1) = \infty \vee |\text{SUCCESSOR}(z, t - 1)| < |x|$ **then**
>>> | $z \leftarrow z + \text{sign}(x)$
>>
>> **end**
>
> **end**
> **if** $x = \text{SUCCESSOR}(z, t - 1)$ **then**
>> | **return** $(z, false)$
>
> **else**
>> ⟨ this can only happen if cell x was generated from $t - 1$ to t ⟩
>> **return** $(z, true)$
>
> **end**

end

Algorithm 3. Counting how many cells existing at time $t - 1$ get indices "closer to 0" than cell z.

function sign: $\mathbb{Z} \to \mathbb{Z}$ with $\text{sign}(0) = 0$, and $\text{sign}(x) = |x|/x$ otherwise, allows one to write down compact code working for both cases as shown in Algorithm 3.

Finally, $\text{SUCCESSOR}(x, t)$ has to be described; see Algorithm 4 on the next page. Roughtly speaking, this is again a simple counting procedure: For every cell "between 0 and up to, but excluding x itself", one simply counts whether it gives rise to 0, 1 or 2 cells at time $t + 1$. After the trivial cases (cell x is deleted, or $x = 0$) have been handled, the consequences of the actions of cell 0 are taken into account. Before the **while** loop is started, i is the number of the first cell $\neq 0$ in the direction of cell x, and z is already the correct value to be returned

---

function *cell z* $\leftarrow$ SUCCESSOR(*cell x*, *time t*) **is**
    if $\sigma(\text{STATE}(x-1,t), \text{STATE}(x,t), \text{STATE}(x+1,t)) = \otimes$ **then return** ∞ **end**

    $\langle$ the rest is only executed if the cell isn't deleted $\rangle$
    if $x = 0$ **then return** 0 **end**

    $\langle$ the rest is only executed if $x > 0$ or $x < 0$ $\rangle$
    if $x > 0$ **then**
        $i \leftarrow 1$
        if $\sigma(\text{STATE}(-1,t), \text{STATE}(0,t), \text{STATE}(1,t)) \neq \otimes$ **then**
            $z \leftarrow 1$
        else
            $z \leftarrow 0$
        end
    end
    if $x < 0$ **then**
        $i \leftarrow -1$
        if $\varepsilon(\text{STATE}(-1,t), \text{STATE}(0,t), \text{STATE}(1,t)) \neq \otimes$ **then**
            $z \leftarrow -2$
        else
            $z \leftarrow -1$
        end
    end
    $\langle$ Now, consider all cells from i up to, but excluding x, $\rangle$
    $\langle$ and count how many non-$\otimes$ cells they amount to $\rangle$
    while $i \neq x$ **do**
        $q_{-1} \leftarrow \text{STATE}(i-1, t)$
        $q_0 \leftarrow \text{STATE}(i, t)$
        $q_1 \leftarrow \text{STATE}(i+1, t)$
        if $\sigma(q_{-1}, q_0, q_1) \neq \otimes$ **then** $z \leftarrow z + \text{sign}(x)$ **end**
        if $\varepsilon(q_{-1}, q_0, q_1) \neq \otimes$ **then** $z \leftarrow z + \text{sign}(x)$ **end**
        $i \leftarrow i + \text{sign}(x)$
    end
    return z
end

Algorithm 4. Counting how many cells are generated from t to $t + 1$ which are "closer to 0" than cell x.

in case none of the cells $i, \ldots, x - \text{sign}(x)$ lead to states $\neq \otimes$. Subsequently, z is increased/decreased (depending on which side of 0 is considered) when required.

Finally, the space complexity of the above algorithm has to be determined depending on the length $n = |w|$ of the input. Let $t(n)$ be the time complexity of the SXCA. First of all, one observes that, inside FINALSTATE, there will be calls to STATE for all t from 0 to $t(n)$. This requires a counter which uses $\log t(n)$ space.

Secondly, the functions STATE, PREDECESSOR and SUCCESSOR all have a parameter time t, and they call themselves recursively in such a way that at least on every second recursion the time parameter is decreased by 1. Hence the number of recursion levels is bounded by $2t(n)$.

It remains to determine the space needed on each recursion level in any of the three functions. In all of them, q a constant number of cell indices have to be stored. Since the number of non-quiescent cells can only grow from k to $2(k+2)$ in a single global step, from an initial configuration with $n \geq 2$ cellscan, after $t(n)$ steps, give rise to at most $n \cdot 4^{t(n)}$ non-quiescent cells. $O(t(n))$ bits suffice for storing the index of any cell in this resulting configuration.

Therefore the total space complexity of the above algorithm is bounded by $O((t(n)^2)$.

5   Conclusion and Outlook

We have proven that space complexity of Turing machines and time complexity of shrinking and expanding cellular automata are polynomially related. In the construction we made use of both shrinking and expanding, but only in a somewhat restricted way: During a first part of the computation the CA was only expanding, and during the second part only shrinking.

It is still an open problem to exactly characterize the set of problems which, for example, can be decided by polynomial time cellular automata which can only expand, but not shrink. The results may depend on the precise definition of acceptance for XCA.

References

1. Chandra, A.K., Kozen, D.C., Stockmeyer, L.J.: Alternation. J. ACM **28**, 114–133 (1981)
2. Fellah, A., Yu, S.: Iterative tree automata, alternating Turing machines, and uniform Boolean circuits: relationships and characterization. In: Berghel, H., et al. (eds.) Proceedings of the 1992 ACM/SIGAPP Symposium on Applied Computing: Technological Challenges of the 1990's, SAC 1992, pp. 1159–1166. ACM, New York (1992)
3. Kutrib, M., Malcher, A., Wendlandt, M.: Shrinking one-way cellular automata. In: Kari, J. (ed.) AUTOMATA 2015. LNCS, vol. 9099, pp. 141–154. Springer, Heidelberg (2015)
4. Modanese, A.C.V.: Shrinking and Expanding Cellular Automata. Bachelor Thesis. Karlsruhe Institute of Technology (2016)

5. Rosenfeld, A., Wu, A., Dubitzki, T.: Fast language acceptance by shrinking cellular automata. Inf. Sci. **30**, 47–53 (1983)
6. van Emde Boas, P.: Machine models and simulations. In: van Leeuwen, J. (ed.) Handbook of Theoretical Computer Science, pp. 1–66. Elsevier (1990)

An 8-State Simple Reversible Triangular Cellular Automaton that Exhibits Complex Behavior

Kenichi Morita$^{(\boxtimes)}$

Hiroshima University, Higashi-hiroshima 739-8527, Japan
km@hiroshima-u.ac.jp

Abstract. A three-neighbor triangular partitioned cellular automaton (TPCA) is a CA whose cell is triangular-shaped and divided into three parts. The next state of a cell is determined by the three adjacent parts of its neighbor cells. The framework of TPCA makes it easy to design reversible triangular CAs. Among them, isotropic 8-state (i.e., each part has two states) TPCAs, which are called elementary TPCAs (ETPCAs), are extremely simple, since each of their local transition functions is described by only four local rules. In this paper, we investigate a specific *reversible* ETPCA T_{0347}, where 0347 is its identification number in the class of 256 ETPCAs. In spite of the simplicity of the local function and the constraint of reversibility, evolutions of configurations in T_{0347} have very rich varieties, and look like those in the Game-of-Life CA to some extent. In particular, a "glider" and "glider guns" exist in T_{0347}. Furthermore, using gliders to represent signals, we can implement universal reversible logic gates in it. By this, computational universality of T_{0347} is derived.

1 Introduction

A three-neighbor triangular cellular automaton (TCA) is a one whose cell is regarded as being triangular-shaped, and communicates with its three neighbor cells. Here, we use the framework of triangular partitioned cellular automata (TPCAs) [6], where each cell is divided into three parts, and each part has its own state set. TPCAs are a subclass of TCAs where the state set of a cell is the Cartesian product of the sets of the three parts. In a TPCA, the next state of a cell is determined depending only on the three adjacent parts of the neighbor cells (not depending on the states of the whole three neighbor cells). Such a framework is useful for designing reversible TCAs.

We define an *elementary* TPCA (ETPCA) as a TPCA such that each part of a cell has only two states (hence a cell has eight states), and its local transition function is isotropic (i.e., rotation-symmetric). There are 256 ETPCAs in total, and there are 36 *reversible* ETPCAs (RETPCAs). ETPCAs are extremely simple, since each of their local functions is described by only four local rules. But, they still show interesting behavior as in the case of one-dimensional elementary

© IFIP International Federation for Information Processing 2016
Published by Springer International Publishing Switzerland 2016. All Rights Reserved
M. Cook and T. Neary (Eds.): AUTOMATA 2016, LNCS 9664, pp. 170–184, 2016.
DOI: 10.1007/978-3-319-39300-1_14

cellular automata (ECAs) [13,14]. In [6], it is shown that the RETPCA with the identification number 0157 (explained later) is computationally universal.

In this paper, we investigate a specific RETPCA T_{0347} having the identification number 0347. It somewhat resembles Game-of-Life CA [2,4,5], and exhibits interesting behavior. In particular, there exist a *glider* and *glider guns*. The glider in T_{0347} is a moving object with period 6. There are glider guns that generate gliders in three directions as well as in one direction. There is also a gun that generates gliders to the negative time direction. We can compose right-turn, U-turn, and left-turn modules out of stable *blocks*, which can change the moving direction of a glider. It is also possible to change the direction of a glider by colliding another glider appropriately. Based on these basic operations, we can implement *gate modules* that simulate reversible logic gates in the cellular space of T_{0347}. By this, computational universality of T_{0347} is concluded.

2 Elementary Triangular Partitioned Cellular Automata

In this section, we give definitions on elementary triangular partitioned cellular automata (ETPCAs), their reversibility, and some related notions.

A *partitioned cellular automaton* (PCA) is a subclass of a standard CA, where a cell is divided into several parts, and each part has its own state set. The next state of a cell is determined by the states of the adjacent parts of the neighboring cells. Figure 1 shows the case of a two-dimensional three-neighbor *triangular PCA* (TPCA). A *local function* of a TPCA is specified by a set of local rules of the form shown in Fig. 1 (b). Applying it to all the cells in parallel, a *global function*, which gives a transition relation among configurations, is obtained. Hereafter, we consider only deterministic PCAs.

We say a PCA is *locally reversible* if its local function is injective, and *globally reversible* if its global function is injective. It is known that global reversibility and local reversibility are equivalent (Lemma 1). Thus, such a PCA is simply called a *reversible PCA* (RPCA). Note that, in [11], the lemma is given for one-dimensional PCAs, but it is easy to extend it for two-dimensional PCAs.

Lemma 1. [11] *A PCA A is globally reversible iff it is locally reversible.*

By this lemma, to obtain a reversible CA, it is sufficient to give a locally reversible PCA. Thus, the framework of PCA makes it easy to design reversible CAs.

A TPCA is called *isotropic* (or *rotation-symmetric*), if, for each local rule, the rules obtained by rotating the both sides of it by a multiple of 60° exist. Note that, if a TPCA is isotropic, then all three parts of a cell must have the same state set. In the following, we study only isotropic TPCAs.

An 8-state isotropic TPCA is called an *elementary TPCA* (ETPCA). Thus, each part of a cell has the state set $\{0,1\}$. ETPCAs are the simplest ones among two-dimensional PCAs. But, this class still contains many interesting PCAs as in the case of one-dimensional elementary CAs (ECAs) [13,14].

Since ETPCA is isotropic, and each part of a cell has only two states, its local function is defined by only four local rules. Hence, an ETPCA can be specified

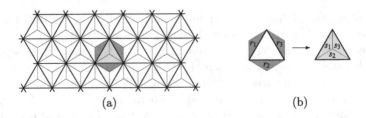

(a)   (b)

Fig. 1. A three-neighbor triangular PCA. (a) Its cellular space, and (b) a local rule.

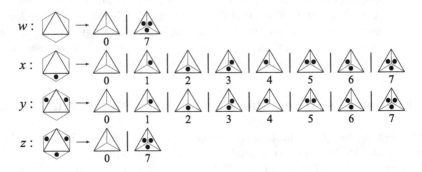

Fig. 2. Representing an ETPCA by a four-digit number $wxyz$, where $w, z \in \{0, 7\}$ and $x, y \in \{0, 1, \ldots, 7\}$. The states 0 and 1 are represented by a blank and $\bullet$, respectively. Vertical bars indicate alternatives of a right-hand side of a rule.

by a four-digit number $wxyz$ such that $w, z \in \{0, 7\}$ and $x, y \in \{0, 1, \ldots, 7\}$ as shown in Fig. 2. Thus, there are 256 ETPCAs. Note that w and z must be 0 or 7 because ETPCAs are isotropic and deterministic. The ETPCA with the identification number $wxyz$ is denoted by T_{wxyz}.

A *reversible ETPCA* is denoted by *RETPCA*. It is easy to see the following.

> An ETPCA T_{wxyz} is reversible iff
> $(w, z) \in \{(0, 7), (7, 0)\} \wedge$
> $(x, y) \in \{1, 2, 4\} \times \{3, 5, 6\} \cup \{3, 5, 6\} \times \{1, 2, 4\}$

Let T_{wxyz} be an ETPCA. We say T_{wxyz} is *conservative* (or *bit-conserving*), if the total number of particles (i.e., $\bullet$'s) is conserved in each rule. Thus, the following holds. Note that, if an ETPCA is conservative, then it is reversible.

> An ETPCA T_{wxyz} is conservative iff
> $w = 0 \wedge x \in \{1, 2, 4\} \wedge y \in \{3, 5, 6\} \wedge z = 7$

3   RETPCA T_{0347} and Its Properties

Here, we focus on a specific RETPCA T_{0347}, and investigate its properties. Its local function is given in Fig. 3. It is a non-conservative RETPCA. We shall see that there are many patterns (i.e., segments of configurations) that show

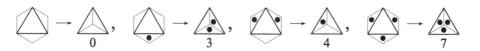

Fig. 3. Local function of the non-conservative RETPCA T_{0347}

interesting behavior as in the case of Game-of-Life CA [2,4,5]. In Sect. 4, some of them will be used to construct reversible logic gates in T_{0347}. In [10], evolving processes of various configurations of T_{0347} can be seen by movies.

3.1   Glider, Block, Fin, and Rotator in T_{0347}

The most useful object in T_{0347} is a *glider* (Fig. 4). It swims in the cellular space like a fish (or an eel). It travels a unit distance, the side-length of a triangle, in 6 steps. By rotating it appropriately, it can move in any of the six directions.

A *block* is an object shown in Fig. 5 (a) or (b). It does not change its pattern if no other object touches it. In this sense, it is a *stable* pattern. There are two kinds of blocks, i.e., type I (Fig. 5 (a)) and type II (Fig. 5 (b)). As explained in Subsect. 3.2, an appropriate type of a block must be used when colliding a glider with it. Combining several blocks, right-turn, U-turn, and left-turn of a glider will be implemented.

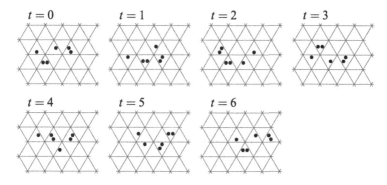

Fig. 4. Movement of a glider. The glider patterns at time $t = 0, \ldots, 5$, and 6 are said to be of phase $0, \ldots, 5$, and 0, respectively.

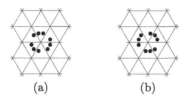

(a)   (b)

Fig. 5. Blocks of (a) type I and (b) type II. They are stable patterns.

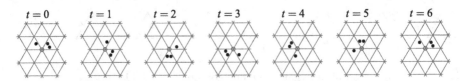

Fig. 6. A fin that rotates around the point indicated by ○

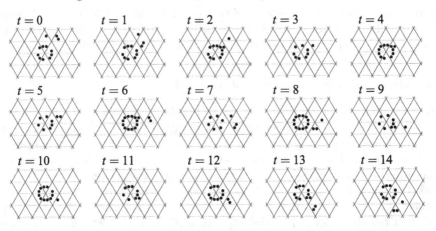

Fig. 7. A fin can travel around a block clockwise. It takes 42 steps to return to the initial position. Note that the block changes its pattern transiently, but it becomes the initial pattern again at $t = 14$.

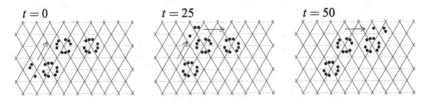

Fig. 8. A fin travelling along a sequence of blocks

A *fin* is an object that simply rotates as in Fig. 6 with period 6. It can also travel around a block (Fig. 7), or a sequence of blocks (Fig. 8) clockwise.

A *rotator* is an object shown in Fig. 9. Like a fin, it rotates around some point, and its period is 42.

3.2   Changing the Move Direction of a Glider by Blocks

We now make several experiments of colliding a glider with blocks. First, we collide a glider with a single block. We put a glider that moves eastward, and a block of type I as shown in Fig. 10 ($t = 0$). At $t = 12$ they collide. Then, the glider is split into a body (i.e., a rotator) and a fin, and the body begins to rotate around the point indicated by ○ ($t = 16$). The fin travels around the block ($t = 38$). When it comes

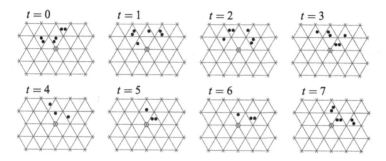

Fig. 9. A rotator that moves around the point indicated by ○. Its period is 42.

back to the original position, it interacts with the body ($t = 50$). By this, the rotation center of the body is shifted upward by two cells, and the fin travels around the block once more ($t = 61$). At $t = 94$, the body and the fin meet again to reconstruct a glider. Finally, the glider goes westward ($t = 97$). By above, backward-turn of the glider is realized.

In the above process, if we use a type II block instead of a type I block, then the glider goes to the north-east direction, but the block cannot be re-used, since some garbage remains (Fig. 11). Therefore, an appropriate type of block should be used depending on the coming direction of the glider. Figure 12 shows the allowed input positions of a glider in each type of a block.

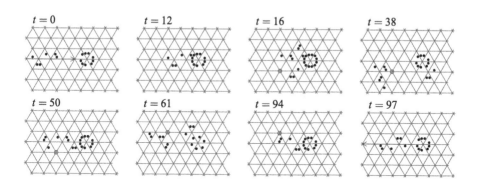

Fig. 10. Colliding a glider with a type I block. It works as a backward-turn module.

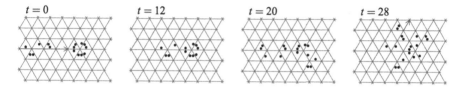

Fig. 11. If we collide a glider moving eastward with a type II block, garbage remains

176   K. Morita

Next, we collide a glider with two blocks (Fig. 13). As in the case of one block (Fig. 10), the glider is split into a rotator and a fin ($t = 56$). The fin travels around the blocks three times without interacting with the rotator. At the end of the fourth round, they meet to form a glider, which goes to the south-west direction ($t = 334$). Hence, two blocks act as a right-turn module.

Figures 14 and 15 show that three blocks and five blocks also act as right-turn modules. They have shorter delays than the case of two blocks. Note that, as in Fig. 15, blocks need not be placed in a straight line. Namely, the sequence of

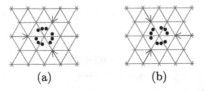

(a)   (b)

Fig. 12. Allowed input positions for (a) the type I block, and (b) the type II block

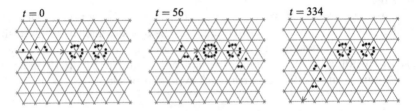

Fig. 13. 120°-right-turn module composed of two blocks

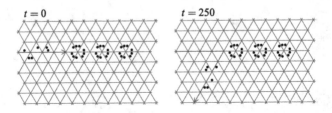

Fig. 14. Right-turn module composed of three blocks

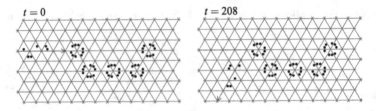

Fig. 15. Right-turn module composed of five blocks, which are not linearly placed

blocks can be bent slightly. But, if we do so, an appropriate type of a block must be used at each position. Otherwise, blocks will be destroyed.

If we collide a glider with a sequence of four, six, or seven blocks, then they will be destroyed. On the other hand, a sequence of eight blocks acts as a backward-turn module like one block, and that of nine blocks acts as a right-turn module like two blocks. Generally, a sequence of $n + 7$ blocks $(n > 0)$ shows a similar behavior as that of n blocks though the total delay is longer. The reason is as follows. The period that the fin goes around the $n+7$ blocks is $36(n+7)+6$. Thus, when the fin comes back, the phase of the rotator becomes the same as in the case of n blocks, since the period of a rotator is 42, and 36×7 is a multiple of 42.

A U-turn module is given in Fig. 16. Also in this case, the glider is first split into a rotator and a fin $(t = 36)$. But, slightly before the fin comes back to the start position, it meets the rotator, and a glider is reconstructed, which moves westward $(t = 113)$. Note that, here the output path is different from the input path, while in the backward-turn module they are the same (Fig. 10).

Figure 17 shows a left-turn module. It is more sophisticated than the right-turn and U-turn modules. The glider is split into a rotator and a fin as before. The fin first travels outside of the module, and then inside. But, around the middle of the module it meets the rotator. A glider is reconstructed from them, and it moves to the north-west direction $(t = 366)$.

It is also possible to make a 60°-right-turn module as in Fig. 18. First, the input glider makes 120°-right-turn by the three blocks $(t = 250)$. Next, the glider is reflected by the left-side block. Finally, it makes 120°-right-turn again by the

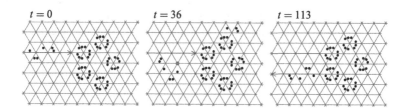

Fig. 16. U-turn module

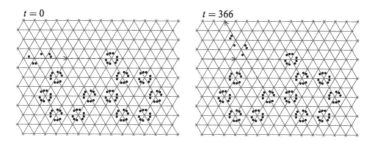

Fig. 17. 120°-left-turn module

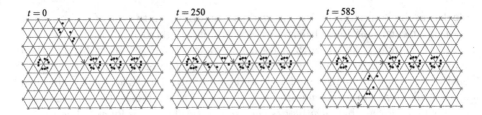

Fig. 18. 60°-right-turn module composed of four blocks

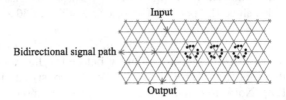

Fig. 19. Interface between a bidirectional signal path, and unidirectional signal paths

three blocks ($t = 585$). This mechanism is also used as an interface between a bidirectional signal path and unidirectional signal paths as shown in Fig. 19.

Table 1 shows the net delay d (i.e., the additional delay caused by the module) and the phase shift s of each turn module. Here, the relation $s = (-d) \bmod 6$ holds. In the case of Fig. 13, we can regard the travelling distance of the glider from $t = 0$ to 334 is 5. Since the speed of a glider is $1/6$, $d = 334 - 5/(1/6) = 304$. Combining these turn modules, we can control the moving direction of a glider freely. It is also possible to shift the phase of a glider. For example, by making right-turn (120°) and then left-turn (120°), its phase is shifted by 2.

Table 1. Net delay and phase shift of the seven turn modules

Module	Net delay d	Phase shift s
Backward-turn	73	+5
Right-turn (120°) by 2 blocks	304	+2
Right-turn (120°) by 3 blocks	220	+2
Right-turn (120°) by 5 blocks	178	+2
U-turn	77	+1
Left-turn (120°)	342	0
Right-turn (60°) by 4 blocks	513	+3

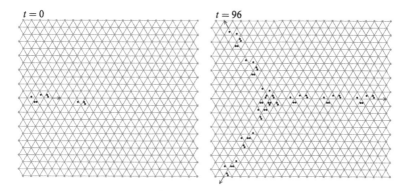

Fig. 20. Three-way glider gun. It generates three gliders every 24 steps.

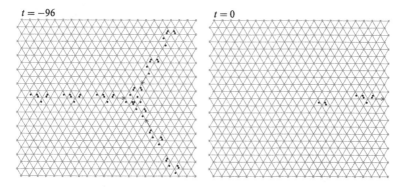

Fig. 21. Glider absorber. It is considered as a glider gun to the negative time direction.

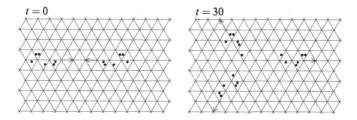

Fig. 22. Generating three gliders by the head-on collision of two gliders

3.3 Glider Guns in T_{0347}

A *glider gun* is a pattern that generates gliders periodically as the one in Game-of-Life [5]. In the RETPCA T_{0347}, it is very easy to create a three-way glider gun. As shown in Fig. 20, it is obtained by colliding a glider with a fin.

Interestingly, there is a *glider absorber* in T_{0347} (Fig. 21). It absorbs three gliders every 24 steps, and finally produces a fin and a glider. It is considered as a "backward glider gun" that generates gliders to the negative time direction.

$t = 0$ Output of glider stream

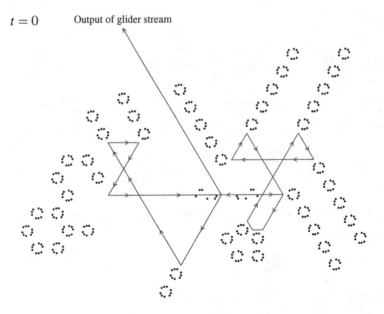

Fig. 23. One-way glider gun. It generates a glider every 1422 steps.

There is another way of composing a glider gun. Figure 22 shows that three gliders are generated by the head-on collision of two gliders. Based on this mechanism, we can design a one-way glider gun as shown in Fig. 23, where two of the three generated gliders are re-used to generate the next three.

4 T_{0347} is Computationally Universal

In this section, we show Turing universality of T_{0347}, i.e., any Turing machine can be simulated in it.

4.1   Showing Turing Universality of Reversible CA

To prove Turing universality of a *reversible* CA, it is sufficient to show that any reversible logic circuit composed of switch gates (Fig. 24 (a)), inverse switch gates (Fig. 24 (b)), and delay elements can be simulated in it (Lemma 6).

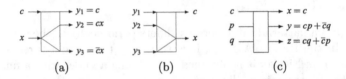

Fig. 24. (a) Switch gate. (b) Inverse switch gate, where $c = y_1$ and $x = y_2 + y_3$ under the assumption $(y_2 \to y_1) \wedge (y_3 \to \overline{y_1})$. (c) Fredkin gate.

Lemma 6 can be derived, e.g., in the following way. First, a Fredkin gate (Fig. 24 (c)) can be constructed out of switch gates and inverse switch gates (Lemma 2). Second, any *reversible sequential machine* (RSM), in particular, a rotary element (RE), which is a 2-state 4-symbol RSM, is composed only of Fredkin gates and delay elements (Lemma 3). Third, any *reversible Turing machine* is constructed out of REs (Lemma 4). Finally, any (irreversible) Turing machine is simulated by a reversible one (Lemma 5). Thus, Lemma 6 follows. Note that the circuit that realizes a reversible Turing machine constructed by this method becomes an infinite (but ultimately periodic) circuit.

Lemma 2. [3] *A Fredkin gate can be simulated by a circuit composed of switch gates and inverse switch gates, which produces no garbage signals.*

Lemma 3. [8] *Any RSM (in particular RE) can be simulated by a circuit composed of Fredkin gates and delay elements, which produces no garbage signals.*

Lemma 4. [9] *Any reversible Turing machine can be simulated by a garbage-less circuit composed only of REs.*

Lemma 5. [1] *Any (irreversible) Turing machine can be simulated by a garbage-less reversible Turing machine.*

Lemma 6. *A reversible CA is Turing universal, if any circuit composed of switch gates, inverse switch gates, and delay elements is simulated in it.*

So far, Turing universality of several reversible two-dimensional CAs has been shown in this way. They are the 2-state reversible block CA model by Margolus [7], the two models of 16-state reversible PCAs on square grid by Morita and Ueno [12], and the conservative RETPCA T_{0157} by Imai and Morita [6].

4.2   Making Switch Gate and Inverse Switch Gate Modules in T_{0347}

We show a switch gate and an inverse switch gate can be implemented in T_{0347} using gliders as signals. The operation of a switch gate is realized by colliding two gliders as shown in Fig. 25. It is important that, in this collision, the glider from the input port c travels to the south-east direction *with no delay* even though it interacts with the glider from x (hence its phase is not shifted also).

Here, we implement a switch gate as a "gate module" in the standard form. Otherwise, adjustment of signal timing, in particular, adjustment of the phase of a glider becomes very cumbersome when designing a larger circuit. A *gate module* is a pattern embedded in a rectangular-like region in the cellular space that satisfies the following (Fig. 26): (1) It realizes a reversible logic gate. (2) Input ports are at the left end. (3) Output ports are at the right end. (4) Delay between input and output is constant and a multiple of 6. Figure 27 shows a switch gate module. The delay in this module is 2232 steps.

An inverse switch gate module is implemented in a similar manner as shown in Fig. 28. It is a "quasi-mirror-image" of the switch gate module. The positions

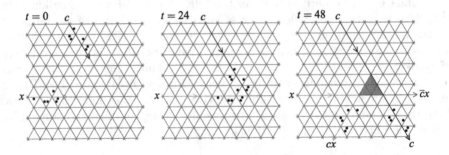

Fig. 25. Switch gate operation realized by collision of two gliders

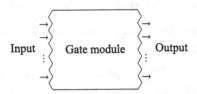

Fig. 26. Gate module in the standard form

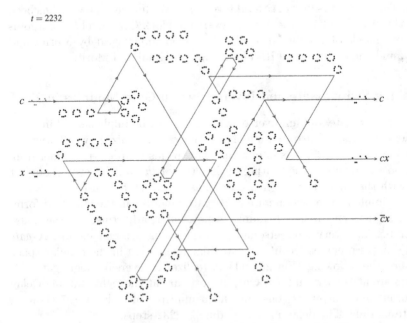

Fig. 27. Switch gate module implemented in T_{0347}

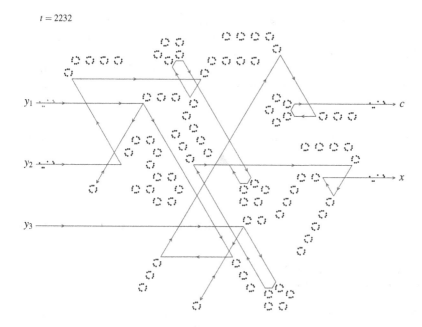

Fig. 28. Inverse switch gate module implemented in T_{0347}

of the blocks are just the mirror images of those in the switch gate. But, each block is replaced by the other type of the corresponding block (not by the mirror image of the block). The delay between input and output is also 2232 steps.

If we use only $120°$-right-turn modules (Table 1) to connect gate modules, then there is no need of adjusting the phases of gliders. This is because the phase shift becomes 0, if we make $120°$-right-turns three times. In this way, we can construct a larger circuit easily. Thus, we have the following theorem.

Theorem 1. *The RETPCA T_{0347} with infinite (but ultimately periodic) config- urations is Turing universal.*

5   Concluding Remarks

Among 256 ETPCAs, we studied a specific reversible ETPCA T_{0347}. In spite of its extreme simplicity of the local function and the constraint of reversibil- ity, T_{0347} shows interesting behavior. Here, a glider plays the key role in T_{0347}. By placing blocks appropriately, trajectory and the phase of a glider can be completely controlled. Logical operation is also performed by interacting gliders. In this way, Turing universality of T_{0347} with infinite configurations was proved.

On the other hand, it is an open problem whether there is a universal con- structor in T_{0347}, which can build any pattern in some specified class of pat- terns (e.g., the class of all patterns consisting of blocks). It is also not known whether universal systems (such as reversible Turing machines, reversible counter machines, or some others) are simulated in the finite configurations of T_{0347}.

Besides T_{0347}, it has already been shown that the conservative RETPCA T_{0157} with infinite configurations is Turing universal [6], where a single particle rather than a glider is used to represent a signal. It is left for the future study to find other ETPCAs that are universal.

Acknowledgement. This work was supported by JSPS KAKENHI Grant Number 15K00019.

References

1. Bennett, C.H.: Logical reversibility of computation. IBM J. Res. Dev. **17**, 525–532 (1973)
2. Berlekamp, E., Conway, J., Guy, R.: Winning Ways for Your Mathematical Plays, vol. 2. Academic Press, New York (1982)
3. Fredkin, E., Toffoli, T.: Conservative logic. Int. J. Theoret. Phys. **21**, 219–253 (1982)
4. Gardner, M.: Mathematical games: The fantastic combinations of John Conway's new solitaire game "life". Sci. Am. **223**(4), 120–123 (1970)
5. Gardner, M.: Mathematical games: On cellular automata, self-reproduction, the Garden of Eden and the game "life". Sci. Am. **224**(2), 112–117 (1971)
6. Imai, K., Morita, K.: A computation-universal two-dimensional 8-state triangular reversible cellular automaton. Theoret. Comput. Sci. **231**, 181–191 (2000)
7. Margolus, N.: Physics-like model of computation. Physica D **10**, 81–95 (1984)
8. Morita, K.: A simple construction method of a reversible finite automaton out of Fredkin gates, and its related problem. Trans. IEICE Japan **E–73**, 978–984 (1990)
9. Morita, K.: A simple universal logic element and cellular automata for reversible computing. In: Margenstern, M., Rogozhin, Y. (eds.) MCU 2001. LNCS, vol. 2055, pp. 102–113. Springer, Heidelberg (2001)
10. Morita, K.: A reversible elementary triangular partitioned cellular automaton that exhibits complex behavior (slides with simulation movies). Hiroshima University Institutional Repository (2016). http://ir.lib.hiroshima-u.ac.jp/00039321
11. Morita, K., Harao, M.: Computation universality of one-dimensional reversible (injective) cellular automata. Trans. IEICE Japan **E72**, 758–762 (1989)
12. Morita, K., Ueno, S.: Computation-universal models of two-dimensional 16-state reversible cellular automata. IEICE Trans. Inf. Syst. **E75–D**, 141–147 (1992)
13. Wolfram, S.: Theory and Applications of Cellular Automata. World Scientific Publishing, Singapore (1986)
14. Wolfram, S.: A New Kind of Science. Wolfram Media Inc, Champaign (2002)

Cellular Automata
on Group Sets and the Uniform
Curtis-Hedlund-Lyndon Theorem

Simon Wacker[✉]

Karlsruhe Institute of Technology, Karlsruhe, Germany
simon.wacker@kit.edu
http://www.kit.edu

Abstract. We introduce cellular automata whose cell spaces are left homogeneous spaces and prove a uniform as well as a topological variant of the Curtis-Hedlund-Lyndon theorem. Examples of left homogeneous spaces are spheres, Euclidean spaces, as well as hyperbolic spaces acted on by isometries; vertex-transitive graphs, in particular, Cayley graphs, acted on by automorphisms; groups acting on themselves by multiplication; and integer lattices acted on by translations.

Keywords: Cellular automata · Group actions · Curtis-Hedlund-Lyndon theorem

In the first chapter of the monograph 'Cellular Automata and Groups' [1], Tullio Ceccherini-Silberstein and Michel Coornaert develop the theory of cellular automata whose cell spaces are groups. Examples of groups are abound: The integer lattices and Euclidean spaces with addition (translation), the one-dimensional unit sphere embedded in the complex plane with complex multiplication (rotation), and the vertices of a Cayley graph with the group structure it encodes (graph automorphisms).

Yet, there are many structured sets that do not admit a structure-preserving group structure. For example: Each Euclidean n-sphere, except for the zero-, one-, and three-dimensional, does not admit a topological group structure; the Petersen graph is not a Cayley graph and does thus not admit an edge-preserving group structure on its vertices. However, these structured sets can be acted on by subgroups of their automorphism group by function application. For example Euclidean n-spheres can be acted on by rotations about the origin and graphs can be acted on by edge-preserving permutations of their vertices.

Moreover, there are structured groups that have more symmetries than can be expressed with the group structure. The integer lattices and the Euclidean spaces with addition, for example, are groups, but addition expresses only their translational symmetries and not their rotational and reflectional ones. Though, they can be acted on by arbitrary subgroups of their symmetry groups, like the ones generated by translations and rotations.

© IFIP International Federation for Information Processing 2016
Published by Springer International Publishing Switzerland 2016. All Rights Reserved
M. Cook and T. Neary (Eds.): AUTOMATA 2016, LNCS 9664, pp. 185–198, 2016.
DOI: 10.1007/978-3-319-39300-1_15

The general notion that encompasses these structure-preserving actions is that of a group set, that is, a set that is acted on by a group. A group set M acted on by G such that for each tuple $(m, m') \in M \times M$ there is a symmetry $g \in G$ that transports m to m' is called *left homogeneous space* and the action of G on M is said to be *transitive*. In particular, groups are left homogeneous spaces — they act on themselves on the left by multiplication.

In this paper, we develop the theory of cellular automata whose cell spaces are left homogeneous spaces which culminates in the proof of a uniform and topological variant of a famous theorem by Morton Landers Curtis, Gustav Arnold Hedlund, and Roger Conant Lyndon from 1969, see the paper 'Endomorphisms and automorphisms of the shift dynamical system' [2]. The development of this theory is greatly inspired by [1].

These cellular automata are defined so that their global transition functions are equivariant under the induced group action on global configurations. Depending on the choice of the cell space, these actions may be plain translations but also rotations and reflections. Exemplary for the first case are integer lattices that are acted on by translations; and for the second case Euclidean n-spheres that are acted on by rotations, but also the two-dimensional integer lattice that is acted on by the group generated by translations and the rotation by $90°$.

Sébastien Moriceau defines and studies a more restricted notion of cellular automata over group sets in his paper 'Cellular Automata on a G-Set' [3]. He requires sets of states and neighbourhoods to be finite. His automata are the global transition functions of, what we call, semi-cellular automata with finite set of states and finite essential neighbourhood.

His automata obtain the next state of a cell by shifting the global configuration such that the cell is moved to the origin, restricting that configuration to the neighbourhood of the origin, and applying the local transition function to that local configuration. Our automata obtain the next state of a cell by determining the neighbours of the cell, observing the states of that neighbours, and applying the local transition function to that observed local configuration. The results are the same but the viewpoints are different, which manifests itself in proofs and constructions.

To determine the neighbourhood of a cell we let the relative neighbourhood semi-act on the right on the cell. That right semi-action is to the left group action what right multiplication is to the corresponding left group multiplication. Many properties of cellular automata are a consequence of the interplay between properties of that semi-action, shifts of global configurations, and rotations of local configurations. That semi-action also plays an important role in our definition of right amenable left group sets, see [4], for which the Garden of Eden theorem holds, see [5], which states that each cellular automaton is surjective if and only if it is pre-injective. For example finitely right generated left homogeneous spaces of sub-exponential growth are right amenable, in particular, quotients of finitely generated groups of sub-exponential growth by finite subgroups acted on by left multiplication.

In Sect. 1 we introduce left group actions and our prime example, which illustrates phenomena that cannot be encountered in groups acting on themselves on the left by multiplication. In Sect. 2 we introduce coordinate systems, cell spaces, and right quotient set semi-actions that are induced by left group actions and coordinate systems. In Sect. 3 we introduce semi-cellular and cellular automata. In Sect. 4 we show that a global transition function does not depend on the choice of coordinate system, is equivariant under the induced left group action on global configurations, is determined by its behaviour in the origin, and that the composition of two global transition functions is a global transition function. In Sect. 5 we prove a uniform and a topological variant of the Curtis-Hedlund-Lyndon theorem, which characterise global transition functions of semi-cellular automata by uniform and topological properties respectively. And in Sect. 6 we characterise invertibility of semi-cellular automata.

1   Left Group Actions

Definition 1. *Let M be a set, let G be a group, let $\triangleright$ be a map from $G \times M$ to M, and let e_G be the neutral element of G. The map $\triangleright$ is called* left group action *of G on M, the group G is said to* act *on M on the left by $\triangleright$, and the triple $(M, G, \triangleright)$ is called* left group set *if and only if*

$$\forall m \in M : e_G \triangleright m = m,$$
$$\forall m \in M \, \forall g \in G \, \forall g' \in G : gg' \triangleright m = g \triangleright (g' \triangleright m).$$

Example 1. Let G be a group. It acts on itself on the left by $\triangleright$ by multiplication.

Example 2. Let M be the Euclidean unit 2-sphere, that is, the surface of the ball of radius 1 in 3-dimensional Euclidean space, and let G be the rotation group. The group G acts on M on the left by $\triangleright$ by function application, that is, by rotation about the origin.

Definition 2. *Let $\triangleright$ be a left group action of G on M and let H be a subgroup of G. The left group action $\triangleright\restriction_{H \times M}$ of H on M is denoted by $\triangleright_H$.*

Definition 3. *Let $\triangleright$ be a left group action of G on M. It is called*

1. transitive *if and only if the set M is non-empty and*

$$\forall m \in M \, \forall m' \in M \, \exists g \in G : g \triangleright m = m';$$

2. free *if and only if*

$$\forall g \in G \, \forall g' \in G : (\exists m \in M : g \triangleright m = g' \triangleright m) \implies g = g'.$$

Example 3. In the situation of Example 1, the left group action is transitive and free.

Example 4. In the situation of Example 2, the left group action is transitive but not free.

Definition 4. *Let $(M, G, \triangleright)$ be a left group set. It is called* left homogeneous space *if and only if the left group action $\triangleright$ is transitive.*

Definition 5. *Let $\triangleright$ be a left group action of G on M, and let m and m' be two elements of M.*

1. *The set $G \triangleright m = \{g \triangleright m \mid g \in G\}$ is called* orbit *of m under $\triangleright$.*
2. *The set $G_m = \{g \in G \mid g \triangleright m = m\}$ is called* stabiliser *of m under $\triangleright$.*
3. *The set $G_{m,m'} = \{g \in G \mid g \triangleright m = m'\}$ is called* transporter *of m to m' under $\triangleright$.*

Example 5. In the situation of Example 3, each orbit is G and each stabiliser is $\{e_G\}$.

Example 6. In the situation of Example 4, for each point $m \in M$, its orbit is M and its stabiliser is the group of rotations about the line through the origin and itself.

Lemma 1. *Let $\triangleright$ be a left group action of G on M, let m and m' be two elements of M that have the same orbit under $\triangleright$, and let g be an element of $G_{m,m'}$. Then, $G_{m'} = g G_m g^{-1}$ and $g G_m = G_{m,m'} = G_{m'} g$.*   □

Definition 6. *Let M and M' be two sets, let f be a map from M to M', and let $\triangleright$ be a left group action of G on M. The map f is called $\triangleright$-invariant if and only if*

$$\forall g \in G \, \forall m \in M : f(g \triangleright m) = f(m).$$

Definition 7. *Let M and M' be two sets, let f be a map from M to M', and let $\triangleright$ and $\triangleright'$ be two left group actions of G on M and M' respectively. The map f is called $(\triangleright, \triangleright')$-equivariant if and only if*

$$\forall g \in G \, \forall m \in M : f(g \triangleright m) = g \triangleright' f(m);$$

and $\triangleright$-equivariant if and only if it is $(\triangleright, \triangleright')$-equivariant, $M = M'$, and $\triangleright = \triangleright'$.

Lemma 2. *Let f be a $(\triangleright, \triangleright')$-equivariant and bijective map from M to M'. The inverse of f is $(\triangleright', \triangleright)$-equivariant*

Lemma 3. *Let G be a group and let H be a subgroup of G. The group G acts transitively on the quotient set G/H on the left by*

$$\cdot \colon G \times G/H \to G/H,$$
$$(g, g'H) \mapsto g g' H.$$

Lemma 4. *Let $\triangleright$ be a transitive left group action of G on M, let m_0 be an element of M, and let G_0 be the stabiliser of m_0 under $\triangleright$. The map*

$$\iota \colon M \to G/G_0,$$
$$m \mapsto G_{m_0, m},$$

is $(\triangleright, \cdot)$-equivariant and bijective.

2   Right Quotient Set Semi-Actions

Definition 8. *Let $\mathcal{M} = (M, G, \rhd)$ be a left homogeneous space, let m_0 be an element of M, let g_{m_0,m_0} be the neutral element of G, and, for each element $m \in M \setminus \{m_0\}$, let $g_{m_0,m}$ be an element of G such that $g_{m_0,m} \rhd m_0 = m$. The tuple $\mathcal{K} = (m_0, \{g_{m_0,m}\}_{m \in M})$ is called* coordinate system *for $\mathcal{M}$; the element m_0 is called* origin; *for each element $m \in M$, the element $g_{m_0,m}$ is called* coordinate *of m; for each subgroup H of G, the stabiliser of the origin m_0 under $\rhd_H$, which is $G_{m_0} \cap H$, is denoted by H_0.*

Definition 9. *Let $\mathcal{M} = (M, G, \rhd)$ be a left homogeneous space and let $\mathcal{K} = (m_0, \{g_{m_0,m}\}_{m \in M})$ be a coordinate system for $\mathcal{M}$. The tuple $\mathcal{R} = (\mathcal{M}, \mathcal{K})$ is called* cell space, *each element $m \in M$ is called* cell, *and each element $g \in G$ is called* symmetry.

Example 7. In the situation of Example 5, let m_0 be the neutral element e_G of G and, for each element $m \in G$, let $g_{m_0,m}$ be the only element in G such that $g_{m_0,m} m_0 = m$, namely m. The tuple $\mathcal{K} = (m_0, \{g_{m_0,m}\}_{m \in M})$ is a coordinate system of $\mathcal{M} = (G, G, \cdot)$ and the tuple $\mathcal{R} = (\mathcal{M}, \mathcal{K})$ is a cell space.

Example 8. In the situation of Example 6, let m_0 be the north pole $(0, 0, 1)^\mathsf{T}$ of M and, for each point $m \in M$, let $g_{m_0,m}$ be a rotation about an axis in the (x, y)-plane that rotates m_0 to m. Note that g_{m_0,m_0} is the trivial rotation. The tuple $\mathcal{K} = (m_0, \{g_{m_0,m}\}_{m \in M})$ is a coordinate system of $\mathcal{M} = (M, G, \rhd)$ and the tuple $\mathcal{R} = (\mathcal{M}, \mathcal{K})$ is a cell space.

In the remainder of this section, let $\mathcal{R} = ((M, G, \rhd), (m_0, \{g_{m_0,m}\}_{m \in M}))$ be a cell space.

Lemma 5. *The map*

$$\lhd \colon M \times G/G_0 \to M,$$
$$(m, gG_0) \mapsto g_{m_0,m} g g_{m_0,m}^{-1} \rhd m \ (= g_{m_0,m} g \rhd m_0),$$

is a right quotient set semi-action *of G/G_0 on M with defect G_0, which means that, for each subgroup H of G such that $\{g_{m_0,m} \mid m \in M\} \subseteq H$,*

$$\forall m \in M : m \lhd G_0 = m,$$
$$\forall m \in M \, \forall h \in H \, \exists h_0 \in H_0 : \forall \mathfrak{g} \in G/G_0 : m \lhd h \cdot \mathfrak{g} = (m \lhd hG_0) \lhd h_0 \cdot \mathfrak{g}.$$

Example 9. In the situation of Example 7, the stabiliser G_0 of the neutral element m_0 under $\cdot$ is the trivial subgroup $\{e_G\}$ of G and, for each element $g \in G$, we have $g_{m_0,m} g g_{m_0,m}^{-1} m = m g m^{-1} m = m g$. Under the natural identification of G/G_0 with G, the induced semi-action $\lhd$ is the right group action of G on itself by multiplication.

Example 10. In the situation of Example 8, the stabiliser G_0 of the north pole m_0 under $\triangleright$ is the group of rotations about the z-axis. An element $gG_0 \in G/G_0$ semi-acts on a point m on the right by the induced semi-action $\triangleleft$ by first rotating m to m_0, $g_{m_0,m}^{-1} \triangleright m = m_0$, secondly rotating m_0 as prescribed by g, $gg_{m_0,m}^{-1} \triangleright m = g \triangleright m_0$, and thirdly undoing the first rotation, $g_{m_0,m}gg_{m_0,m}^{-1} \triangleright m = g_{m_0,m} \triangleright (g \triangleright m_0)$, in other words, by first changing the rotation axis of g such that the new axis stands to the line through the origin and m as the old one stood to the line through the origin and m_0, $g_{m_0,m}gg_{m_0,m}^{-1}$, and secondly rotating m as prescribed by this new rotation.

Let N_0 be a subset of the sphere M, which we think of as a geometrical object on the sphere that has its centre at m_0, for example, a circle of latitude. The set $N = \{gG_0 \in G/G_0 \mid g \triangleright m_0 \in N_0\} = \{G_{m_0,m} \mid m \in N_0\} = \{g_{m_0,m}G_0 \mid m \in N_0\}$ can be thought of as a realisation of N_0 in G/G_0. Indeed, $m_0 \triangleleft N = g_{m_0,m_0}\{g_{m_0,m} \mid m \in N_0\} \triangleright m_0 = N_0$. Furthermore, for each point $m \in M$, the set $m \triangleleft N = g_{m_0,m} \triangleright N_0$ has the same shape and size as N_0 but its centre at m.

Lemma 6. *The semi-action $\triangleleft$ is*

1. transitive, *which means that the set M is non-empty and*

$$\forall m \in M \, \forall m' \in M \, \exists \mathfrak{g} \in G/G_0 : m \triangleleft \mathfrak{g} = m';$$

2. free, *which means that*

$$\forall \mathfrak{g} \in G/G_0 \, \forall \mathfrak{g}' \in G/G_0 : (\exists m \in M : m \triangleleft \mathfrak{g} = m \triangleleft \mathfrak{g}') \implies \mathfrak{g} = \mathfrak{g}'.$$

Lemma 7. *The semi-action $\triangleleft$*

1. semi-commutes with $\triangleright$, *which means that, for each subgroup H of G such that $\{g_{m_0,m} \mid m \in M\} \subseteq H$,*

$$\forall m \in M \, \forall h \in H \, \exists h_0 \in H_0 : \forall \mathfrak{g} \in G/G_0 : (h \triangleright m) \triangleleft \mathfrak{g} = h \triangleright (m \triangleleft h_0 \cdot \mathfrak{g});$$

2. exhausts its defect with respect to its semi-commutativity with $\triangleright$ in m_0, *which means that, for each subgroup H of G such that $\{g_{m_0,m} \mid m \in M\} \subseteq H$,*

$$\forall h_0 \in H_0 \, \forall \mathfrak{g} \in G/G_0 : (h_0^{-1} \triangleright m_0) \triangleleft \mathfrak{g} = h_0^{-1} \triangleright (m_0 \triangleleft h_0 \cdot \mathfrak{g}).$$

Lemma 8. *The maps*

$$\left\{ \begin{array}{l} m_0 \triangleleft \_\colon G/G_0 \to M, \\ \qquad\qquad \mathfrak{g} \mapsto m_0 \triangleleft \mathfrak{g}, \end{array} \right\} \quad and \quad \left\{ \begin{array}{l} \iota\colon M \to G/G_0, \\ \quad m \mapsto G_{m_0,m}, \end{array} \right\}$$

are inverse to each other and, under the identification of G/G_0 with M by either of these maps,

$$\forall m \in M \, \forall \mathfrak{g} \in G/G_0 \simeq M : m \triangleleft \mathfrak{g} = g_{m_0,m} \triangleright \mathfrak{g}.$$

3 Semi-Cellular and Cellular Automata

In this section, let $\mathcal{R} = (\mathcal{M}, \mathcal{K}) = ((M, G, \triangleright), (m_0, \{g_{m_0, m}\}_{m \in M}))$ be a cell space.

Definition 10. *Let Q be a set, let N be a subset of G/G_0 such that $G_0 \cdot N \subseteq N$, and let δ be a map from Q^N to Q. The quadruple $\mathcal{C} = (\mathcal{R}, Q, N, \delta)$ is called* semi-cellular automaton, *each element $q \in Q$ is called* state, *the set N is called* neighbourhood, *each element $n \in N$ is called* neighbour, *and the map δ is called* local transition function.

Example 11. In the situation of Example 9, the semi-cellular automata over $\mathcal{R}$ are the usual cellular automata over the group G.

Example 12. In the situation of Example 10, let Q be the set $\{0, 1\}$, let N_0 be the union of all circles of latitude between $45°$ and $90°$ north, which is a curved circular disk of radius $\pi/4$ with the north pole m_0 at its centre, let N be the set $\{gG_0 \mid g \triangleright m_0 \in N_0\}$, and let

$$\delta \colon Q^N \to Q, \ell \mapsto \begin{cases} 0, & \text{if } \forall n \in N : \ell(n) = 0, \\ 1, & \text{if } \exists n \in N : \ell(n) = 1. \end{cases}$$

The quadruple $\mathcal{C} = (\mathcal{R}, Q, N, \delta)$ is a semi-cellular automaton.

In the remainder of this section, let $\mathcal{C} = (\mathcal{R}, Q, N, \delta)$ be a semi-cellular automaton.

Definition 11. *Let E be a subset of N. It is called* essential neighbourhood *if and only if*

$$\forall \ell \in Q^N \, \forall \ell' \in Q^N : \ell \!\restriction_E = \ell' \!\restriction_E \implies \delta(\ell) = \delta(\ell').$$

Definition 12. *Each map $\ell \in Q^N$ is called* local configuration. *The stabiliser G_0 acts on Q^N on the left by*

$$\bullet \colon G_0 \times Q^N \to Q^N,$$
$$(g_0, \ell) \mapsto [n \mapsto \ell(g_0^{-1} \cdot n)].$$

Definition 13. *The semi-cellular automaton $\mathcal{C}$ is called* cellular automaton *if and only if its local transition function δ is $\bullet$-invariant.*

Example 13. In the situation of Example 11, the stabiliser G_0 of the neutral element m_0 is the trivial subgroup $\{e_G\}$ of G. Therefore, for each semi-cellular automaton over $\mathcal{R}$, its local transition function is $\bullet$-invariant and hence it is a cellular automaton.

Example 14. In the situation of Example 12, think of 0 as black, 1 as white, and of local configurations as black-and-white patterns on $N_0 = m_0 \triangleleft N$. The rotations G_0 about the z-axis act on these patterns by $\bullet$ by rotating them. The local transition function δ maps the black pattern to 0 and all others to 1, which is invariant under rotations. Therefore, the quadruple $\mathcal{C}$ is a cellular automaton.

Definition 14. *Each map* $c \in Q^M$ *is called* global configuration. *The group* G
acts on Q^M *on the left by*

$$\blacktriangleright : G \times Q^M \to Q^M,$$

$$(g, c) \mapsto [m \mapsto c(g^{-1} \rhd m)].$$

Definition 15. *For each global configuration* $c \in Q^M$ *and each cell* $m \in M$, *the*
local configuration

$$N \to Q,$$

$$n \mapsto c(m \lessdot n),$$

is called observed by m in c.

Remark 1. Because the semi-action $\lessdot$ is free, for each local configuration $\ell \in Q^N$
and each cell $m \in M$, there is a global configuration $c \in Q^M$ such that the local
configuration observed by m in c is ℓ.

Definition 16. *The map*

$$\Delta : Q^M \to Q^M,$$

$$c \mapsto [m \mapsto \delta(n \mapsto c(m \lessdot n))],$$

is called global transition function.

Example 15. In the situation of Example 14, repeated applications of the global
transition function of $\mathcal{C}$ grows white regions on M.

Remark 2. For each subset A of M and each global configuration $c \in Q^M$, the
states of the cells A in $\Delta(c)$ depends at most on the states of the cells $A \lessdot N$ in
c. More precisely,

$$\forall A \subseteq M \, \forall c \in Q^M \, \forall c' \in Q^M : c \restriction_{A \lessdot N} = c' \restriction_{A \lessdot N} \implies \Delta(c) \restriction_A = \Delta(c') \restriction_A .$$

Lemma 9. *Let* m *be an element of* M, *let* g *be an element of* G, *and let* g_0 *be*
an element of G_0 *such that*

$$\forall n \in N : (g^{-1} \rhd m) \lessdot n = g^{-1} \rhd (m \lessdot g_0 \cdot n).$$

For each global configuration $c \in Q^M$,

$$[n \mapsto c((g^{-1} \rhd m) \lessdot n)] = g_0^{-1} \bullet [n \mapsto (g \blacktriangleright c)(m \lessdot n)].$$

Definition 17. *The set* $N_0 = m_0 \lessdot N$ *is called* neighbourhood *of* m_0.

Definition 18. *The map*

$$\delta_0 : Q^{N_0} \to Q,$$

$$\ell_0 \mapsto \delta(n \mapsto \ell_0(m_0 \lessdot n)),$$

is called local transition function *of* m_0.

Lemma 10. *The global transition function* Δ *of* $\mathcal{C}$ *is identical to the map*

$$\Delta_0 : Q^M \to Q^M,$$

$$c \mapsto [m \mapsto \delta_0((g_{m_0,m}^{-1} \blacktriangleright c) \restriction_{N_0})].$$

4 Invariance, Equivariance, Determination, and Composition of Global Transition Functions

In Theorem 1 we show that a global transition function does not depend on the choice of coordinate system. In Theorem 2 we show that a global transition function is ►-equivariant if and only if the local transition function is •-invariant. In Theorem 3 we show that a global transition function is determined by its behaviour in the origin. And in Theorem 4 we show that the composition of two global transition functions is a global transition function.

Lemma 11. *Let* $\rhd$ *be a left group action of* G *on* M. *The group* G *acts on* $\bigcup_{m \in M} G/G_m$ *on the left by*

$$\circ : G \times \bigcup_{m \in M} G/G_m \to \bigcup_{m \in M} G/G_m,$$
$$(g, g'G_m) \mapsto gg'G_m g^{-1} \ (= gg'g^{-1}G_{g \rhd m}),$$

such that, for each element $g \in G$ *and each element* $m \in M$, *the map*

$$(g \circ \_) \restriction_{G/G_m \to G/G_{g \rhd m}} : G/G_m \to G/G_{g \rhd m},$$
$$g'G_m \mapsto g \circ g'G_m,$$

is bijective.

Lemma 12. *Let* $\mathcal{M} = (M, G, \rhd)$ *be a left homogeneous space, let* $\mathcal{K} = (m_0, \{g_{m_0,m}\}_{m \in M})$ *and* $\mathcal{K}' = (m_0', \{g_{m_0',m}'\}_{m \in M})$ *be two coordinate systems for* $\mathcal{M}$, *let* H *be a subgroup of* G *such that* $\{g_{m_0,m}\}|_{m \in M} \cup \{g_{m_0',m}'\}|_{m \in M} \subseteq H$, *and let* h *be an element of* H *such that* $h \rhd m_0 = m_0'$. *Then,*

$$\forall m \in M \, \exists h_0 \in H_0 : \forall \mathfrak{g}' \in G/G_0' : m \vartriangleleft' \mathfrak{g}' = m \vartriangleleft h_0 \cdot (h^{-1} \circ \mathfrak{g}').$$

Theorem 1. *In the situation of Lemma 12, let* $\mathcal{C} = ((\mathcal{M}, \mathcal{K}), Q, N, \delta)$ *be a semi-cellular automaton such that* δ *is* $\bullet_{H_0}$-*invariant, let* N' *be the set* $h \circ N$, *and let*

$$\delta' : Q^{N'} \to Q,$$
$$\ell' \mapsto \delta(n \mapsto \ell'(h \circ n)).$$

The quadruple $((\mathcal{M}, \mathcal{K}'), Q, N', \delta')$ *is a semi-cellular automaton whose global transition function is identical to the one of* $\mathcal{C}$.

Corollary 1. *Let* $(\mathcal{M}, \mathcal{K}) = ((M, G, \rhd), (m_0, \{g_{m_0,m}\}_{m \in M}))$ *be a cell space and let* $\mathcal{C} = ((\mathcal{M}, \mathcal{K}), Q, N, \delta)$ *be a cellular automaton. For each coordinate system* $\mathcal{K}' = (m_0', \{g_{m_0',m}'\}_{m \in M})$ *for* $\mathcal{M}$, *there is a cellular automaton* $((\mathcal{M}, \mathcal{K}'), Q, N', \delta')$ *whose global transition function is identical to the one of* $\mathcal{C}$. □

Theorem 2. *Let* $\mathcal{R} = ((M, G, \rhd), (m_0, \{g_{m_0,m}\}_{m \in M}))$ *be a cell space, let* $\mathcal{C} = (\mathcal{R}, Q, N, \delta)$ *be a semi-cellular automaton, and let* H *be a subgroup of* G *such that* $\{g_{m_0,m} \mid m \in M\} \subseteq H$.

1. *If the local transition function δ is $\bullet_{H_0}$-invariant, then the global transition function Δ is $\blacktriangleright_H$-equivariant.*
2. *If there is an $\blacktriangleright_H$-equivariant map $\Delta_0 \colon Q^M \to Q^M$ such that*

$$\forall c \in Q^M : \Delta_0(c)(m_0) = \delta(n \mapsto c(m_0 \lessdot n)), \tag{1}$$

   then the local transition function δ is $\bullet_{H_0}$-invariant.
3. *The local transition function δ is $\bullet_{H_0}$-invariant if and only if the global transition function Δ is $\blacktriangleright_H$-equivariant.*

Corollary 2. *Let C be a semi-cellular automaton. It is a cellular automaton if and only if its global transition function is $\blacktriangleright$-equivariant.* □

Theorem 3. *Let $\mathcal{R} = ((M, G, \rhd), (m_0, \{g_{m_0,m}\}_{m \in M}))$ be a cell space, let $C = (\mathcal{R}, Q, N, \delta)$ be a semi-cellular automaton, let Δ_0 be a map from Q^M to Q^M, and let H be a subgroup of G such that $\{g_{m_0,m} \mid m \in M\} \subseteq H$. The following statements are equivalent:*

1. *The local transition function δ is $\bullet_{H_0}$-invariant and the global transition function of C is Δ_0;*
2. *The global transition function Δ_0 is $\blacktriangleright_H$-equivariant and*

$$\forall c \in Q^M : \Delta_0(c)(m_0) = \delta(n \mapsto c(m_0 \lessdot n)). \tag{2}$$

Theorem 4. *Let $\mathcal{R} = ((M, G, \rhd), (m_0, \{g_{m_0,m}\}_{m \in M}))$ be a cell space, let $C = (\mathcal{R}, Q, N, \delta)$ and $C' = (\mathcal{R}, Q, N', \delta')$ be two semi-cellular automata, and let H be a subgroup of G such that $\{g_{m_0,m} \mid m \in M\} \subseteq H$, and δ and δ' are $\bullet_{H_0}$-invariant. Furthermore, let*

$$N'' = \{g \cdot n' \mid n \in N, n' \in N', g \in n\}$$

and let

$$\delta'' \colon Q^{N''} \to Q,$$
$$\ell'' \mapsto \delta(n \mapsto \delta'(n' \mapsto \ell''(g_{m_0, m_0 \lessdot n} \cdot n'))).$$

The quadruple $C'' = (\mathcal{R}, Q, N'', \delta'')$ is a semi-cellular automaton whose local transition function is $\bullet_{H_0}$-invariant and whose global transition function is $\Delta \circ \Delta'$.

5   Curtis-Hedlund-Lyndon Theorems

In this section we equip the the set of global configurations with a uniform and a topological structure, and prove a uniform and a topological variant of the Curtis-Hedlund-Lyndon theorem. In Main Theorem 5, the uniform variant, we show that global transition functions are characterised by $\blacktriangleright$-equivariance and uniform continuity. And in its Corollary 3, the topological variant, that under the assumption that the set of states is finite they are characterised by $\blacktriangleright$-equivariance and continuity.

Definition 19. *Let G be a group equipped with a topology. It is called* topolog-ical *if and only if the maps*

$$\left\{\begin{array}{l} G \times G \to G, \\ (g, g') \mapsto gg', \end{array}\right\} \ and \ \left\{\begin{array}{l} G \to G, \\ g \mapsto g^{-1}, \end{array}\right\}$$

are continuous, where $G \times G$ is equipped with the product topology.

Definition 20. *Let M be a set, let $\mathfrak{L}$ be a subset of the power set of M, and let T be a subset of M. The set T is called* transversal *of $\mathfrak{L}$ if and only if there is a surjective map $f\colon \mathfrak{L} \to T$ such that for each set $A \in \mathfrak{L}$ we have $f(A) \in A$.*

Definition 21. *Let M and M' be two topological spaces and let f be a contin-uous map from M to M'. The map f is called*

1. proper *if and only if, for each compact subset K of M', its preimage $f^{-1}(K)$ is a compact subset of M;*
2. semi-proper *if and only if, for each compact subset K of M', each transversal of $\{f^{-1}(k) \mid k \in K\}$ is included in a compact subset of M.*

Definition 22. *Let M be a topological space, let G be a topological group, and let $\mathcal{M} = (M, G, \triangleright)$ be a left group set. The group set $\mathcal{M}$ is called*

1. topological *if and only if the map $\triangleright$ is continuous;*
2. proper *if and only if it is topological and the so-called* action map

$$\alpha\colon G \times M \to M \times M,$$
$$(g, m) \mapsto (g \triangleright m, m),$$

is proper, where $G \times M$ and $M \times M$ are equipped with their respective product topology;
3. semi-proper *if and only if it is topological and its action map is semi-proper.*

Remark 3. Each proper map is semi-proper and each proper left group set is semi-proper.

Lemma 13. *Let $\mathcal{M} = (M, G, \triangleright)$ be a semi-proper left group set, and let K and K' be two compact subsets of M. Each transversal of $\{G_{k',k} \mid (k, k') \in K \times K'\}$ is included in a compact subset of G.*

Lemma 14. *Let $\mathcal{M} = (M, G, \triangleright)$ be a left group set. Equip M and G with their respective discrete topology. The group set $\mathcal{M}$ is semi-proper.*

Definition 23. *Let $\mathcal{R} = ((M, G, \triangleright), (m_0, \{g_{m_0,m}\}_{m \in M}))$ be a topological or uni-form cell space. Equip G/G_0 with the topology or uniformity induced by $m_0 \triangleleft \_$.*

Lemma 15. *Let $\mathcal{R} = ((M, G, \triangleright), (m_0, \{g_{m_0,m}\}_{m \in M}))$ be a semi-proper cell space, let K be a compact subset of M, and let E be a compact subset of G/ G_0. The set $K \triangleleft E$ is included in a compact subset of M.*

Definition 24. *Let M be a topological space and let Q be a set.*

1. *The topology on Q^M that has for a subbase (and base) the sets*

$$\mathfrak{E}(K,b) = \{c \in Q^M \mid c \restriction_K = b\}, \text{ for } b \in Q^K \text{ and } K \subseteq M \text{ compact,}$$

is called topology of discrete convergence on compacta.
2. *The uniformity on Q^M that has for a subbase (and base) the sets*

$$\mathfrak{E}(K) = \{(c,c') \in Q^M \times Q^M \mid c \restriction_K = c' \restriction_K\}, \text{ for } K \subseteq M \text{ compact,}$$

is called uniformity of discrete convergence on compacta.

Remark 4. 1. Let M be equipped with the discrete topology. For each subset A of M, the set A is compact if and only if it is finite. Therefore, the topology and uniformity of discrete convergence on compacta on Q^M are the prodiscrete topology and uniformity on Q^M respectively. For definitions of the latter see Sects. 1.2 and 1.9 in [1]. The set Q^M equipped wit the prodiscrete topology is Hausdorff (see Proposition 1.2.1 in [1]) and if Q is finite then it is compact (see the first paragraph in Sect. 1.8 in [1]).
2. If the topological space M is compact, the topology and uniformity of discrete convergence on compacta on Q^M is the discrete topology and uniformity on Q^M respectively.
3. The topology induced by the uniformity of discrete convergence on compacta on Q^M is the topology of discrete convergence on compacta on Q^M.

Remark 5. Let $\mathcal{R} = ((M,G,\triangleright), (m_0, \{g_{m_0,m}\}_{m \in M}))$ be a cell space and let Q be a finite set. Equip M and G with their respective discrete topologies, and equip Q^M with the prodiscrete topology. According to Lemma 14, the cell space $\mathcal{R}$ is semi-proper. Because the topology on G/G_0 is discrete, each subset E of G/G_0 is compact if and only if it is finite. Because Q^M is compact, each map $\Delta: Q^M \to Q^M$ is uniformly continuous if and only if it is continuous. And, because Q^M is Hausdorff, each map $\Delta: Q^M \to Q^M$, is a uniform isomorphism if and only if it is continuous and bijective.

Main Theorem 5 (Uniform Variant; Morton Landers Curtis, Gustav Arnold Hedlund, and Roger Conant Lyndon, 1969). *Let $\mathcal{R} = ((M,G,\triangleright), (m_0, \{g_{m_0,m}\}_{m \in M}))$ be a semi-proper cell space, let Q be a set, let Δ be a map from Q^M to Q^M, let Q^M be equipped with the uniformity of discrete convergence on compacta, and let H be a subgroup of G such that $\{g_{m_0,m} \mid m \in M\} \subseteq H$. The following statements are equivalent:*

1. *The map Δ is the global transition function of a semi-cellular automaton with $\bullet_{H_0}$-invariant local transition function and compact essential neighbourhood.*
2. *The map Δ is $\blacktriangleright_H$-equivariant and uniformly continuous.*

Proof. First, let Δ be the global transition function of a semi-cellular automaton $\mathcal{C} = (\mathcal{R}, Q, N, \delta)$ such that δ is $\bullet_{H_0}$-invariant and such that there is a compact essential neighbourhood E of $\mathcal{C}$. Then, according to Item 3 of Theorem 2, the map Δ is $\blacktriangleright_H$-equivariant. Moreover, let K be a compact subset of M. According to Lemma 15, the set $K \vartriangleleft E$ is included in a compact subset L of M. For each $c \in Q^M$ and each $c' \in Q^M$, if $c \restriction_{K \vartriangleleft E} = c' \restriction_{K \vartriangleleft E}$, then $\Delta(c) \restriction_K = \Delta(c') \restriction_K$, in particular, if $c \restriction_L = c' \restriction_L$, then $\Delta(c) \restriction_K = \Delta(c') \restriction_K$. Thus,

$$(\Delta \times \Delta)(\mathfrak{E}(L)) \subseteq \mathfrak{E}(K).$$

Because the sets $\mathfrak{E}(K)$, for $K \subseteq M$ compact, constitute a base of the uniformity on Q^M, the global transition function Δ is uniformly continuous.

Secondly, let Δ be as in Item 2. Because Δ is uniformly continuous, there is a compact subset E_0 of M such that

$$(\Delta \times \Delta)(\mathfrak{E}(E_0)) \subseteq \mathfrak{E}(\{m_0\}).$$

Therefore, for each $c \in Q^M$, the state $\Delta(c)(m_0)$ depends at most on $c \restriction_{E_0}$. The subset $E = (m_0 \vartriangleleft \_)^{-1}(E_0) = \{G_{m_0,m} \mid m \in E_0\}$ of G/G_0 is compact. Let N be the set $G_0 \cdot E$. Then, $G_0 \cdot N \subseteq N$. And, because $E_0 \subseteq m_0 \vartriangleleft N$, for each $c \in Q^M$, the state $\Delta(c)(m_0)$ depends at most on $c \restriction_{m_0 \vartriangleleft N}$. Hence, there is a map $\delta : Q^N \to Q$ such that

$$\forall c \in Q^M : \Delta(c)(m_0) = \delta(n \mapsto c(m_0 \vartriangleleft n)).$$

The quadruple $\mathcal{C} = (\mathcal{R}, Q, N, \delta)$ is a semi-cellular automaton. Conclude with Theorem 3 that δ is $\bullet_{H_0}$-invariant and that Δ is the global transition function of $\mathcal{C}$. $\square$

Corollary 3 (Topological Variant; Morton Landers Curtis, Gustav Arnold Hedlund, and Roger Conant Lyndon, 1969). *Let $\mathcal{R} = ((M, G, \vartriangleright), (m_0, \{g_{m_0,m}\}_{m \in M}))$ be a cell space, let Q be a finite set, let Δ be a map from Q^M to Q^M, let Q^M be equipped with the prodiscrete topology, and let H be a subgroup of G such that $\{g_{m_0,m} \mid m \in M\} \subseteq H$. The following statements are equivalent:*

1. *The map Δ is the global transition function of a semi-cellular automaton with $\bullet_{H_0}$-invariant local transition function and finite essential neighbourhood.*
2. *The map Δ is $\blacktriangleright_H$-equivariant and continuous.*

Proof. With Remark 5 this follows directly from Main Theorem 5. $\square$

Remark 6. In the case that $M = G$ and $\vartriangleright$ is the group multiplication of G, Main Theorem 5 is Theorem 1.9.1 in [1] and Corollary 3 is Theorem 1.8.1 in [1].

6   Invertibility

Definition 25. *Let $\mathcal{C} = (\mathcal{R}, Q, N, \delta)$ be a semi-cellular automaton. It is called invertible if and only if there is a semi-cellular automaton $\mathcal{C}'$, called inverse to $\mathcal{C}$, such that the global transition functions of $\mathcal{C}$ and $\mathcal{C}'$ are inverse to each other.*

Theorem 6. *Let* $\mathcal{R} = ((M, G, \triangleright), (m_0, \{g_{m_0,m}\}_{m \in M}))$ *be a semi-proper cell space, let Q be a set, let Δ be a map from Q^M to Q^M, let Q^M be equipped with the uniformity of discrete convergence on compacta, and let H be a subgroup of G such that $\{g_{m_0,m} \mid m \in M\} \subseteq H$. The following statements are equivalent:*

1. *The map Δ is the global transition function of an invertible semi-cellular automaton $\mathcal{C}$ that has an inverse $\mathcal{C}'$ such that the local transition functions of $\mathcal{C}$ and $\mathcal{C}'$ are $\bullet_{H_0}$-invariant, and $\mathcal{C}$ and $\mathcal{C}'$ have compact essential neighbourhoods.*
2. *The map Δ is an $\blacktriangleright_H$-equivariant uniform isomorphism.*

Proof. With Lemma 2 this follows directly from Main Theorem 5.

Corollary 4. *Let $\mathcal{R} = ((M, G, \triangleright), (m_0, \{g_{m_0,m}\}_{m \in M}))$ be a cell space, let Q be a finite set, let Δ be a map from Q^M to Q^M, let Q^M be equipped with the prodiscrete topology, and let H be a subgroup of G such that $\{g_{m_0,m} \mid m \in M\} \subseteq H$. The following statements are equivalent:*

1. *The map Δ is the global transition function of an invertible semi-cellular automaton $\mathcal{C}$ that has an inverse $\mathcal{C}'$ such that the local transition functions of $\mathcal{C}$ and $\mathcal{C}'$ are $\bullet_{H_0}$-invariant, and $\mathcal{C}$ and $\mathcal{C}'$ have finite essential neighbourhoods.*
2. *The map Δ is $\blacktriangleright_H$-equivariant, continuous, and bijective.*

Proof. With Remark 5 this follows directly from Theorem 6. $\square$

Corollary 5. *Let $\mathcal{R} = ((M, G, \triangleright), (m_0, \{g_{m_0,m}\}_{m \in M}))$ be a cell space, let H be a subgroup of G such that $\{g_{m_0,m} \mid m \in M\} \subseteq H$. Furthermore, let $\mathcal{C} = (\mathcal{R}, Q, N, \delta)$ be a semi-cellular automaton with finite set of states, finite essential neighbourhood, and $\bullet_{H_0}$-invariant local transition function. The automaton $\mathcal{C}$ is invertible if and only if its global transition function is bijective.*

Proof. With Item 1 of Theorem 2 and Corollary 3 this follows directly from Corollary 4. $\square$

References

1. Ceccherini-Silberstein, T., Coornaert, M.: Cellular Automata and Groups. Springer Monographs in Mathematics. Springer, Heidelberg (2010)
2. Hedlund, G.A.: Endomorphisms and automorphisms of the shift dynamical system. Math. Syst. Theory **3**(4), 320–375 (1969)
3. Moriceau, S.: Cellular Automata on a G-Set. J. Cell. Automata **6**(6), 461–486 (2011)
4. Wacker, S.: Right Amenable Left Group Sets and the Tarski-Følner Theorem. arXiv: 1603.06460 [math.GR]
5. Wacker, S.: The Garden of Eden Theorem for Cellular Automata on Group Sets. arXiv: 1603.07272 [math.GR]

Author Index

Printed in the United States
By Bookmasters